INLAND
NAVIGATION SYSTEM PLANNING

The Upper Mississippi River—Illinois Waterway

Committee to Review the Upper Mississippi River-Illinois Waterway
Navigation System Feasibility Study

Water Science and Technology Board
Division on Earth and Life Studies

Transportation Research Board

National Research Council

NATIONAL ACADEMY PRESS
Washington, D.C.

NATIONAL ACADEMY PRESS • 2101 Constitution Avenue, N.W. • Washington, D.C. 20418

NOTICE: The project that is the subject of this report was approved by the Governing Board of the National Research Council, whose members are drawn from the councils of the National Academy of Sciences, the National Academy of Engineering, and the Institute of Medicine. The members of the committee responsible for the report were chosen for their special competencies and with regard for appropriate balance.

Support for this project was provided by the Department of the Army under Contract No. DASW01-00-C-3017.

International Standard Book Number 0-309-07405-3

Inland Navigation System Planning: The Upper Mississippi River-Illinois Waterway is available from the National Academy Press, 2101 Constitution Avenue, N.W., Washington, D.C. 20418, (800) 624-6242 or (202) 334-3313 (in the Washington metropolitan area); internet <http://www.nap.edu>.

Cover design by Van Nguyen, National Academy Press. Photos courtesy of Jon Duyvejonck, U.S. Fish and Wildlife Service. Photo on back cover, Bridge with Moon, J&D Richardson Stock Photography, Memphis, TN, http://www.jdrichardson.com/memphis1.html.

Printed in the United States of America

The **National Academy of Sciences** is a private, nonprofit, self-perpetuating society of distinguished scholars engaged in scientific and engineering research, dedicated to the furtherance of science and technology and to their use for the general welfare. Upon the authority of the charter granted to it by the Congress in 1863, the Academy has a mandate that requires it to advise the federal government on scientific and technical matters. Dr. Bruce M. Alberts is president of the National Academy of Sciences.

The **National Academy of Engineering** was established in 1964, under the charter of the National Academy of Sciences, as a parallel organization of outstanding engineers. It is autonomous in its administration and in the selection of its members, sharing with the National Academy of Sciences the responsibility for advising the federal government. The National Academy of Engineering also sponsors engineering programs aimed at meeting national needs, encourages education and research, and recognizes the superior achievement of engineers. Dr. William A. Wulf is president of the National Academy of Engineering.

The **Institute of Medicine** was established in 1970 by the National Academy of Sciences to secure the services of eminent members of appropriate professions in the examination of policy matters pertaining to the health of the public. The Institute acts under the responsibility given to the National Academy of Sciences by its congressional charter to be an adviser to the federal government and, upon its own initiative, to identify issues of medical care, research, and education. Dr. Kenneth I. Shine is president of the Institute of Medicine.

The **National Research Council** was organized by the National Academy of Sciences in 1916 to associate the broad community of science and technology with the Academy's purposes of furthering knowledge and advising the federal government. Functioning in accordance with general policies determined by the Academy, the Council has become the principal operating agency of both the National Academy of Sciences and the National Academy of Engineering in providing services to the government, the public, and the scientific and engineering communities. The Council is administered jointly by both Academies and the Institute of Medicine. Dr. Bruce M. Alberts and Dr. William A. Wulf are chairman and vice chairman, respectively, of the National Research Council.

COMMITTEE TO REVIEW THE UPPER MISSISSIPPI RIVER-ILLINOIS WATERWAY NAVIGATION SYSTEM FEASIBILITY STUDY

LESTER B. LAVE, *Chair*, Carnegie Mellon University, Pittsburgh, Pennsylvania
PHILLIP BAUMEL, Iowa State University, Ames
KENNETH D. BOYER, Michigan State University, Ann Arbor
MICHAEL S. BRONZINI, George Mason University, Fairfax, Virginia
KENNETH L. CASAVANT, Washington State University, Pullman
BONNIE G. COLBY, University of Arizona, Tucson
JONATHAN P. DEASON, George Washington University, Washington, D.C.
JOSÉ A. GÓMEZ-IBÁNEZ, Harvard University, Cambridge, Massachusetts
DELON HAMPTON, Delon Hampton & Associates, Chartered, Washington, D.C.
EDWIN E. HERRICKS, University of Illinois, Urbana
DAVID H. MOREAU, University of North Carolina, Chapel Hill

NRC Staff

JEFFREY W. JACOBS, Study Director, Water Science and Technology Board
ANITA A. HALL, Senior Project Assistant, Water Science and Technology Board
JOSEPH MORRIS, Senior Staff Officer, Transportation Research Board
SUSAN GARBINI, Senior Staff Officer, Transportation Research Board

WATER SCIENCE AND TECHNOLOBY BOARD

Staff

Preface

In its study of the prospects for extending several locks on the Upper Mississippi River-Illinois Waterway (UMR-IWW) system, the U.S. Army Corps of Engineers found itself in the middle of a clash of powerful forces and changing values. The traditional argument has been that the federal government must do everything it can to open the continent to people, promote commerce, and aid exporters. This view continues to be supported by some members of Congress, farmers and tow companies from the areas surrounding these rivers, as well as the Port of New Orleans and exporters. Holders of this view support improvements to navigation system infrastructure on the UMR–IWW in order to carry grain to New Orleans for export. They contend that if tows are now twice as long as in the past, the U.S. should extend the locks to relieve congestion and improve the competitiveness of grain farmers. Complicating this view is a rapidly changing world in which the U.S. has gone from a position of feeding much of the world after World War II to a position of competing, not always successfully, against low cost grain exports from South America and other places. The market for American grain over the next five years is highly uncertain. The market over the next century is unknowable.

Meanwhile, those promoting environmental values now argue that improving navigation must not needlessly damage recreation and river ecology. These same interests also seek reversal of some of the profound changes in river ecology brought about by transforming a free flowing river into a series of pools. Most Americans gave little attention to environmental pollution in 1960; certainly few knew or cared about changes in river ecology. Many environmental views that were extreme in 1970 have become mainstream views in 2001.

In this setting, the Corps was charged with analyzing the benefits and costs of a $1 billion improvement in transportation infrastructure, the extensions of several locks on the UMR–IWW. The benefits arise from the demand for U.S. grain in other nations. Forecasting the demand for even the next few years is difficult. Costs include construction and operations. These costs could be increased greatly by insisting that new infrastructure changes begin to reverse the changes in river ecology caused by the past construction of the locks and dams and the passage of tows. Indeed, the benefit of the infrastructure change would be reduced to zero at the time that the public demanded that the dams be breached to allow a free flowing river.

While the mission of this committee never made us the focus of the contending parties, our two meetings in which we asked for public input made the controversy abundantly clear to us. At a narrow, technical level, we pored over the Corps' analyses of the price sensitivity of the current and future demand for barge shipment, the effects of additional traffic on congestion and thus on shipping costs, and the analysis of the environmental effects of construction and the passage of tows. The complexity of these tasks would test any government agency. Accomplishing them in the midst of heated controversy makes them all the more difficult.

No member of the committee is "on the line" in the same way that Corps analysts are. Most of us have the luxury of examining issues at length and without contending parties scrutinizing each step. Thus, we appreciate the difficulty of the Corps' task. Nonetheless, we found that substantial improvements can be made in the Corps' analysis that will help inform decision-makers about the implications of this and other waterway projects. We urge Congress, the Department of the Army, and the Corps to give careful attention to improving the analyses so that society can be better informed about the implications of alternative courses of action.

The committee's study was conducted with appreciation of the analytical and political challenges facing the Corps. We greatly appreciate the Corps of Engineers' cooperation and assistance in this study and would especially like to thank Phillip Anderson, Ken Barr, Jim Johnson, Gary Loss, Denny Lundberg, Rich Manguno, Dave Tipple, and Hans van Winkle. We also thank Joseph Westphal and James Smyth from the office of the Assistant Secretary of the Army for Civil Works for their cooperation and advice. Thanks are also due to Chris Brescia of MARC 2000 (St. Louis), George Dusenberry of the Northeast-Midwest Center (Washington, D.C.), Jon Duyvejonck of the U.S. Fish and Wildlife Service (Rock Island, Illinois), Ken Lubinski of the U.S. Geological Survey (LaCrosse, Wisconsin), and Tim Searchinger of Environmental Defense (Washington, D.C.), all of whom spoke with our committee. We are also grateful to the many citizens and interest group representatives who spoke with and educated our committee. We also thank the Corps of Engineers' staff in the St. Louis district office for hosting an informative and enjoyable mid-August field trip to locks and dams 25 and 26. The committee could not have completed its task without the able assistance of NRC staff members Steve Parker, Steve Godwin, Joe Morris, Susan Garbini, and Anita Hall; and Leonard Shabman (Virginia Polytechnic Institute and State University) who was working in the office of the Water Science and Technology Board as a visiting scholar during the conduct of our study. Rhonda Bitterli provided excellent editorial advice and assistance. We are especially indebted to Jeffrey Jacobs, who served as our chief staff officer and helped us craft the language that said what we wanted to say.

Lester B. Lave
Chairman

Acknowledgment of Reviewers

This report was reviewed in draft form by individuals chosen for their diverse perspectives and technical expertise, in accordance with procedures approved by the NRC's Report Review Committee. The purpose of this independent review is to provide candid and critical comments that will assist the institution in making its published report as sound as possible and to ensure that the report meets institutional standards for objectivity, evidence, and responsiveness to the study charge. The review comments and draft manuscript remain confidential to protect the integrity of the deliberative process. We wish to thank the following individuals for their reviews of this report:

Gregory B. Baecher, University of Maryland
John Cairns, Jr., Virginia Polytechnic Institute and State University
Jeanne N. Clarke, University of Arizona
Stephen W. Fuller, Texas A&M University
Gerald E. Galloway, Jr., International Joint Commission
John R. Meyer, Harvard University, Emeritus
Kenneth A. Small, University of California, Irvine

Although the reviewers listed above have provided many constructive comments and suggestions, they were not asked to endorse the conclusions or recommendations nor did they see the final draft of the report before its release. The review of this report was overseen by Debra Knopman, Progressive Policy Institute, appointed by Division on Earth and Life Studies and Sherwin Rosen, The University of Chicago, appointed by the NRC's Report Review Committee who were responsible for making certain that an independent examination of this report was carried out in accordance with institutional procedures and that all review comments were carefully considered. Responsibility for the final content of this report rests entirely with the authoring committee and the institution.

Contents

APPENDIXES

Executive Summary

In 1988, the U.S. Army Corps of Engineers began an investigation of the benefits and costs of extending several locks on the lower portion of the Upper Mississippi River–Illinois Waterway (UMR–IWW) in order to relieve increasing waterway congestion, particularly for grain moving to New Orleans for export. With passage of the Flood Control Act of 1936, Congress required that the Corps conduct a benefit-cost analysis as part of its water resources project planning; Congress will fund water resources projects only if a project's benefits exceed its costs. As economic analysis generally, and benefit-cost analysis in particular, has become more sophisticated, and as environmental and social considerations and analysis have become more important, Corps planning studies have grown in size and complexity. The difficulty in commensurating market and nonmarket costs and benefits also presents the Corps with a significant challenge. The Corps' analysis of the UMR–IWW has extended over a decade, has cost roughly $50 million, and has involved consultations with other federal agencies, state conservation agencies, and local citizens. The analysis has included many consultants and has produced dozens of reports.

In February 2000, the Department of the Army requested that the National Academies review the Corps' final feasibility report. After discussions and negotiations with the Army, in April 2000 the National Academies launched this review and appointed an expert committee to carry it out. The committee, appointed under the joint auspices of the Academies' Water Science and Technology Board (WSTB) and Transportation Research Board (TRB), conducted its study in accordance with the following statement of task:

> This study will focus on the U.S. Army Corps of Engineers' economic analysis regarding proposed improvements, including economic assumptions, methods, and forecasts regarding barge transportation demand on the Upper Mississippi River-Illinois Waterway. The Corps must also consider larger water resources project planning issues, such as formal U.S. federal water resource planning guidelines, possible environmental impacts, and the costs of navigation improvements. Thus, while the committee will focus on the Corps' economic analysis, they will also comment upon the extent to which these larger issues are being appropriately considered in the navigation system feasibility study.

In September 2000, the Corps reported that completion of its feasibility study would be delayed. As the committee was contractually obligated to finish its study in February 2001, it was clear that a final Corps study would not be available for the committee's review. Nonetheless, the National Academies and the Department of the Army agreed that an interim report would be useful. The committee's report thus focuses on a July 2000 draft of the feasibility study, on a draft environmental impact statement, and on numerous studies and reports that the Corps presented to the committee as key supporting documentation in the draft report.

The committee is mindful of its original assignment to examine the Corps' final report. As the Corps has yet to finish its feasibility study, the assessment and recommendations in Chapter 4 do *not* critique the Corps' final report, but rather comment upon a draft of the final report and relevant supporting documents provided by the Corps. Because the Corps decided that more analysis was required before it could produce a final feasibility report, the Corps may have already changed some of the plans the committee describes and analyzes in Chapter 4. The committee thus presents its report with the hope that it will prove useful to the Corps as it continues the analysis and preparation of the final feasibility report.

The prospective lock extensions on the lower UMR–IWW are highly controversial, as was made clear by the high level of participation and passion displayed by proponents and opponents during the committee's meetings and public comment sessions in Washington, D.C. and St. Louis in June and August 2000, respectively. It is not easy to pursue this difficult analysis of the UMR–IWW in such a highly charged atmosphere.

The Corps attempted to build the first comprehensive model of grain use and exports from the area surrounding the river system, and attempted to build a model of the environmental effects of extending the locks and increasing barge traffic. These two studies are exceedingly difficult, and the committee commends the Corps for undertaking them and making important advances. The committee also offers recommendations on how the final feasibility study could be improved over the draft reviewed in this report.

This report's recommendations for improving navigation system planning on the Upper Mississippi and Illinois system are divided into four areas: economics, inland waterway and water resources system planning, the environment, and engineering. The Corps itself may not be able to unilaterally implement all of the report's recommendations. The Corps operates at the behest of the U.S. Congress, and the committee recognized the roles of Congress in granting enabling authorities and appropriations for Corps water resources studies and projects. In addition, the federal Office of Management and Budget (OMB) establishes fiscal priorities for water resources projects. The Corps' discretion to carry out certain studies and projects is thus not always matched by the funds to do so.

The Corps is, however, granted a certain degree of latitude in various authorizations and in its feasibility studies. The committee thus primarily directed recommendations that can be implemented within the Corps' discretion to the Corps of Engineers. Recommendations beyond the Corps' discretion to enact are directed to the U.S. Congress.

Some of the report's recommendations will require sustained and significant resources in order to be implemented, requiring Congressional appropriations. The Department of Defense's request for a rapid evaluation of the draft study meant that some potentially useful aspects of the committee's review could not be accomplished in such a short time frame. The Army agreed that the committee should not be asked to estimate the resources required to implement the committee's recommendations. Nonetheless, the committee was cognizant of the limited nature of available funds. We discussed how limited resources should be managed to provide the most valuable information to the public, Corps, and Congress to inform their decisions concerning river and waterway management decisions and policies.

ECONOMICS

Spatial Equilibrium Model and ESSENCE

The Corps developed a theoretical spatial equilibrium model for the UMR–IWW feasibility study to help forecast future levels of barge traffic across the entire navigation system. This system model represents a major advance over previous economics models used by the Corps to forecast barge traffic. That this spatial equilibrium model should be used as a foundation for the feasibility study. The Corps also developed the ESSENCE model, which calculates equilibrium values for barge traffic and waterway congestion and calculates changes in barge shipping costs that are consistent with waterway traffic forecasts and with past delay patterns at locks. The ESSENCE model does not, however, adequately use the more important concepts of the spatial equilibrium model that were advocated in the draft feasibility study.

Despite advances represented in the spatial equilibrium model and the ESSENCE model, many of the assumptions and data used as input to these models are flawed. Forecasts of future global grain demand did not adequately account for global or domestic supply and demand factors. In some instances, simple linear extrapolations were used in constructing demand forecasts, a practice unlikely to produce satisfactory results. Assumptions regarding the sensitivity and variability of barge shipping rates were not empirically sound.

As a result of flawed assumptions and data, the current (September 2000) results of the spatial equilibrium model and the ESSENCE model should not be used in the feasibility study. The problem lies not in the theoretical motivation behind these models, but in their implementation and data used as input. To correct these problems, the Corps should: (1) obtain a satisfactory database of grain and other relevant freight shipments by barge and alternative modes that includes quantity, origin and destination, and price, (2) revise the ESSENCE model, eliminating assumptions that shipment costs are proportional to distance and that agricultural yields are uniform, (3) estimate demand and supply sensitivities to price from studies of current data, and assure that model parameters reflect these price sensitivities, and (4) include spatial equilibrium concepts in its ESSENCE model.

Managing Congestion

The locks and dams on the UMR–IWW are presently not being used efficiently. Shippers and tow operators bear needlessly high costs because there is no traffic management system for the waterway. Shipments are delayed and costs are driven up by tows that sometimes must wait many hours to pass through a lock, while at other times, locks sit idle. If barge traffic was distributed more evenly, congestion would decrease and shipping costs would fall. Rather than waiting a decade for relief from the congestion by extending the locks, shippers and towboat operators could enjoy immediate improvements through better traffic management.

Only a narrow range of alternatives for addressing waterway congestion on the UMR–IWW was assessed in the feasibility study. Several relatively inexpensive, nonstructural options exist for reducing UMR–IWW traffic congestion, including better scheduling, tradable lockage permits, and congestion fees. Furthermore, it is not clear how the benefits of lock extensions can be evaluated adequately without first managing waterway traffic more efficiently on the existing system. **Congress should instruct the Corps to explore fully these nonstructural options for improving traffic management as the baseline condition for the National Economic Development alternative and environmental evaluation of any proposal for lock extensions. A comprehensive review and assessment of the benefits and costs of nonstructural options for improving traffic management should be conducted. The benefits and costs of lock extensions should not be calculated until nonstructural measures for waterway traffic management have been carefully assessed.**

INLAND WATERWAY AND WATER RESOURCES SYSTEM PLANNING

Integrated Systems Planning

The feasibility study lacks a comprehensive assessment of how changes in navigation might affect economic, environmental, and social systems. For example, the study does not describe the relations between the river's environmental resources and the substantial economies (tourism and recreation) that depend on those resources. Environmental effects of changes in barge traffic have economic implications, but these are not considered in the feasibility study.

A thorough analysis that supports informed decisions must address environmental impacts with the same comprehensiveness and sophistication that is now expected for the evaluation of the National Economic Development alternative. **The Corps should aim toward a more comprehensive and integrated assessment of the navigation system's effects on the environment in the UMR–IWW.**

The Corps has conducted many environmental investigations as part of the feasibility study. However, it is not clear how these environmental studies are incorporated into the decision regarding lock extensions on the Upper Mississippi. **The Corps should clarify the nature of the relations between environmental studies and the decision-making process regarding proposed lock extensions.**

Independent Review

The feasibility study is exceedingly complex, spanning more than a decade and costing over $50 million. The ability to adequately conduct such an extensive study would test the capabilities of most federal agencies. The Corps' study is too important to not receive an independent judgment on the merits of the various approaches and a careful scrutiny of the analysis. The feasibility study would benefit from a second opinion from an independent, expert, and interdisciplinary body from outside the Corps of Engineers and Department of Defense. **Congress**

should thus direct the Corps to have the waterway system management and lock extension feasibility study reviewed by an interdisciplinary group of experts—including environmental and social scientists—from outside the Corps of Engineers.

U.S. Federal *Principles and Guidelines*

The water resources project planning framework and criteria contained in the *Principles and Guidelines* are generally satisfactory and offer the Corps adequate flexibility to undertake comprehensive water resources and navigation system investigations. Corps of Engineers Headquarters has issued clarifying guidance that allows for such planning to proceed. For example, Corps district offices now have instructions to incorporate public values, especially in terms of serving both environmental and National Economic Development objectives, when defining planning problems and opportunities and formulating alternative plans. In addition, the Corps is expected to utilize modern analytical techniques such as system transportation modeling. However, for reasons not clear to this committee, this comprehensive planning framework is not reflected in the draft feasibility study. **The Secretary of the Army should ensure that the environmental consequences of proposed construction and operating practices be analyzed along with the National Economic Development account. Furthermore, environmental improvements—not just the mitigation of incremental environmental damages—should be examined as part of the navigation feasibility study.**

ENVIRONMENT

Environmental concerns have become increasingly important in the last three decades. Large-scale infrastructure changes last a long time, long enough for human and social perceptions and preferences to change significantly. Environmental groups and some government agencies have criticized the feasibility study, charging that is has not treated environmental issues and resources adequately. In the committee's judgment, environmental concerns are likely to become even more important in the future. Environmental concerns have become a core issue in the operation of inland waterway systems and should be treated as such in planning studies. The Corps should recognize the central importance of environmental issues and adapt its planning, engineering design, operations, and analysis accordingly.

Effects of the Navigation System on the Environment

The construction of dozens of locks and dams on the Upper Mississippi and Illinois rivers, the creation of a series of huge navigation pools, wing dams and other river-training structures, and barge traffic have had numerous and complex environmental effects. Although there has been some systematic research into the environmental effects of human and social activities on the Upper Mississippi River (the federal Environmental Management Program being a good example), the understanding of the complex ecosystem dynamics in the UMR–IWW system is limited in many areas. Despite numerous environmental assessments conducted as part of the feasibility study, characterization of the current environmental system is insufficient, as it is in the early stages of scientific validation. The combination of past construction and continuing operations and maintenance continue to affect the river's environmental systems. Gaps in current scientific understanding make it very difficult to accurately understand how additional changes will affect the river. Environmental studies on systemwide effects, cumulative effects, and site-specific effects are needed. **Systemwide research should be conducted on the following topics in the UMR–IWW: (1) cumulative effects of the existing navigation system on river ecology, (2) environmental effects of recent navigation system improvements, (3) cumulative effects of increased towboat passage, and (4) site-specific effects of future construction activities on the UMR–IWW.**

Congress should provide the resources for this research, but the responsibility for conducting this research not be shouldered by the Corps alone; relevant federal and state agencies should participate with the Corps in this research. Since the late 1980s, the Corps has participated in a federal, multi-agency environmental research program in the Upper Mississippi River—the Environmental Management Program (EMP). The EMP, with support of its participating agencies, has gathered much fundamental data on environmental changes in the Upper Mississippi basin. **Congress should continue to provide support for EMP-based research on the links between the navigation system and river ecology. The EMP research effort should be enhanced to improve assessment of the current navigation system's cumulative effects on the environment, and broadened to include studies of the impacts of barge traffic on river ecology.**

Adaptive Management

Adaptive management is a planning approach that strives to meet environmental, economic, and social objectives while keeping management policies flexible and adjustable. An adaptive management approach would be useful to the Corps in addressing and responding to the great environmental and economic uncertainties that attend UMR–IWW system planning. Components of an adaptive management strategy would include ecosystem monitoring, adjusting management policies in light of new information, recognizing the values of ecosystem resiliency, variability, and diversity, recognizing the limits of the river's resources, and seeking to avoid irreversible decisions. **The Corps should seek the authority and funding from Congress necessary to conduct its navigation feasibility study based on the principles of adaptive management that have been articulated in the natural resources management literature.**

Adaptive Mitigation Strategy

The draft feasibility study describes a strategy for mitigating the environmental effects of incremental increases in barge traffic from lock extensions and from the site-specific impacts of construction. The draft study explains a strategy for monitoring the results of mitigation activities, and subsequent adjustments to those mitigation activities, described as an "adaptive mitigation" strategy. The steps in this approach are reasonable. But on a broader scale, this adaptive mitigation strategy is insufficient. The environmental effects of additional barge traffic and construction are not known with a reasonable degree of precision. Consequently, the extent to which these effects might be mitigated, and the associated costs, are also not known. The Corps' approach to considering environmental resources only after locks have been extended is inadequate. In this approach, the environment is treated as a planning constraint rather than as a resource on par with waterway infrastructure investments. Moreover, the process by which adaptive mitigation measures would be identified, implemented, evaluated, and consequently adjusted, is not described. This process should include relevant federal and state agencies, as well as representation of the public. **The Corps' adaptive mitigation strategy is inconsistent with the principles of adaptive management articulated in the natural resources management literature.**

ENGINEERING

If the UMR–IWW locks are extended, the costs of rehabilitating the existing, aging locks, would be greatly reduced. The Corps estimates that the savings through these reduced rehabilitation costs would be considerable. The committee reviewed the basis for these savings and their magnitude. The Corps' modeling approach is sensible; however, benefits should be recalculated based upon new waterway traffic demand forecasts.

During the feasibility study, the Corps revised the contingency estimates for the costs of extending the locks. The committee recognizes the Corps' expertise in designing and building structures such as locks and dams. The Corps proposes a novel method for lock extension. Although the Corps has not used the method, it has been used extensively. As the Corps gains experience with this method in modifying the first few locks, they should be able to estimate future costs more accurately and to find means to lower costs. However, many factors have escalated lock and dam costs in the past and it seems prudent to expect that construction costs might increase significantly due to a variety of factors. **A 25 percent cost contingency is likely to be too low, particularly since recent experience with Lock 26 suggests that major escalation of costs can occur.**

1

Introduction

The Upper Mississippi River–Illinois Waterway (UMR–IWW) is an important component of the U.S. inland navigation system. Only the Lower Mississippi River and the Ohio River carry more barge traffic. Commercial barge traffic on the Upper Mississippi carries grain from the U.S. Corn Belt region downstream to New Orleans, where it is then shipped to international markets. Other commercial cargo shipped on the Upper Mississippi is mainly fertilizer (upstream), coal (both upstream and downstream), and petroleum. In 1995, 126 million tons of cargo were shipped on the UMR–IWW (USACE, 1995).

The central feature of this navigation system is 29 locks and dams on the Upper Mississippi River that help maintain a permanent 9-foot channel for barge traffic (Figure 1.1). These locks and dams were constructed and are operated by the U.S. Army Corps of Engineers (USACE, or Corps). In addition to these locks and dams, the Corps is responsible for the construction, operation, and maintenance of a variety of river channel training structures and activities on the UMR–IWW, including wing dams, revetments, and dredging.

In 1989, the Corps initiated investigations to determine the economic viability of extending locks on the lower portion of the Upper Mississippi River. Many of the locks and dams were constructed in the 1930s and were originally designed for tows up to 600 feet long. Since then, the length of a typical tow has doubled, and the volume of river traffic has increased significantly. Increases in the congestion of river traffic resulted in calls from the shipping and grain industries for the extension of locks to help alleviate the congestion. The Corps initiated the feasibility study of the system to determine if the extension of several locks on the lower portion of the UMR–IWW would be economically justifiable.

In an inland navigation study, the Corps must consider a variety of economic, engineering, and environmental factors; sophisticated analytical techniques are used to estimate demand, cost, and environmental consequences. The Corps typically conducts its water resources project planning studies for individual, site-specific water resources projects. This study, however, considered the entire Upper Mississippi River–Illinois Waterway navigation system and is among the larger system-wide studies the Corps has conducted.

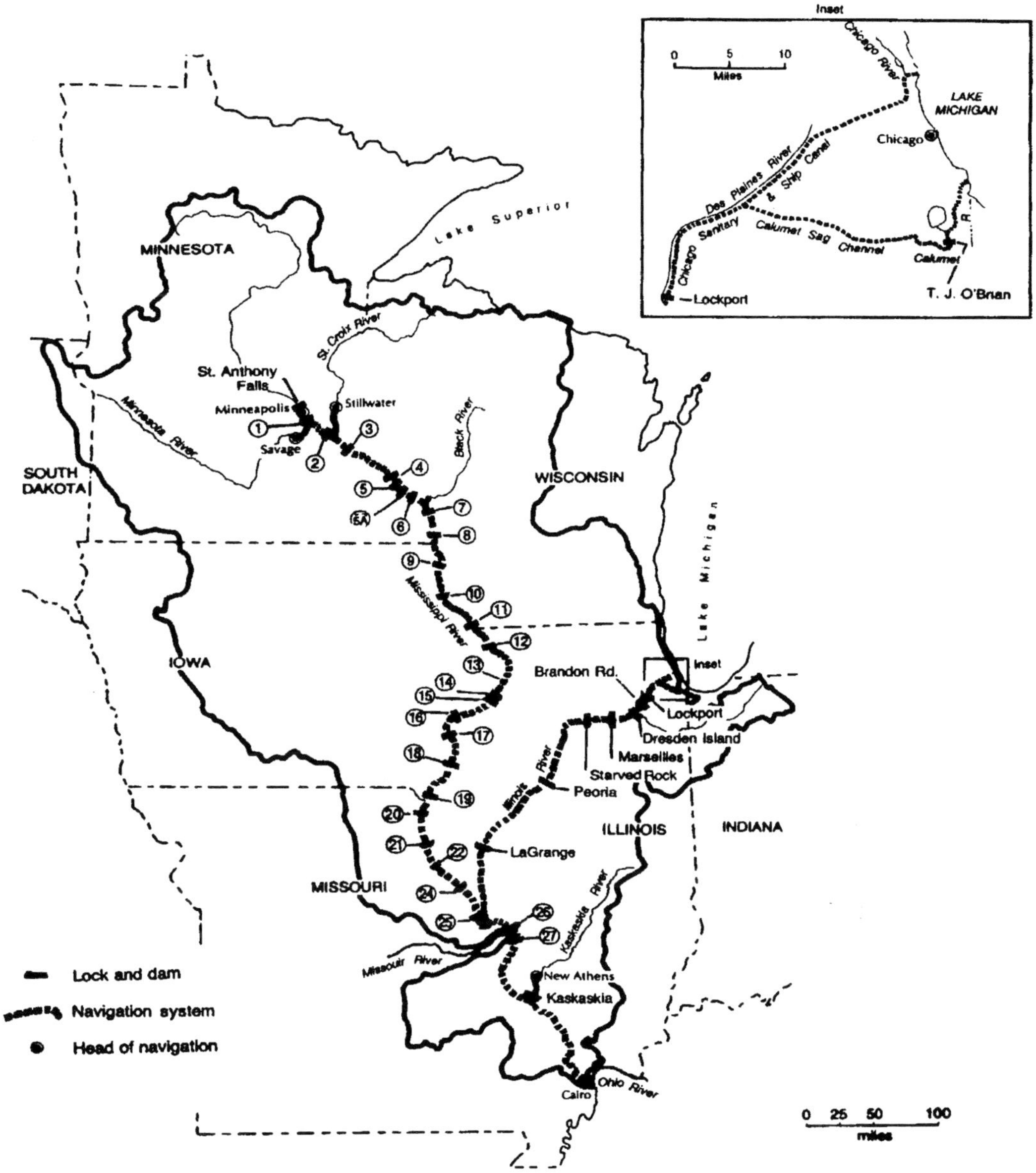

FIGURE 1.1 Locks and Dams on the Upper Mississippi River–Illinois Waterway System. SOURCE: USGS, 1999.

Before the study was completed, controversies arose over the Corps' assumptions and methods employed in the economics portion of the study. To assess the integrity of the Corps' study, the U.S. Department of Defense (DOD) requested the National Academies to conduct a review of the Corps' Upper Mississippi River–Illinois Waterway system navigation feasibility study.

UPPER MISSISSIPPI RIVER NAVIGATION

The Corps' navigation-enhancement activities on the Upper Mississippi River date back to the nineteenth century. In 1864, the U.S. Congress authorized the first system-wide project to improve the river for navigation by providing a 4-foot channel. Because of limitations in project funding and scope, the Corps made few changes to the river, and navigation continued to be hindered by river snags, overhanging trees, and a shifting river channel. In 1878, Congress authorized a 4.5-foot navigation channel project for the Upper Mississippi. This was during a period of significant growth in population and in agricultural production in the region, and the navigation project was initiated partly in response to the threat of a bulk shipping monopoly by the railroads. From the 1870s to the first decade of the twentieth century, timber (especially white pine) was the most important commodity moved on the Upper Mississippi (Anfinson, 1996). By roughly 1910, however, the importance of timber fell as the white pine forests of western Wisconsin and northern Minnesota were depleted, leaving the Upper Mississippi River with no important commodity. The early twentieth century also saw the decline of steamboat traffic on the river, and railroads were carrying a larger share of grain and passenger traffic.

Nonetheless, Congress authorized a 6-foot navigation channel project in 1907, and the Corps began building more wing dams and adding to the existing ones. These wing dams, which extend into the river from the main shoreline or from the bank of an island, were intended to narrow the river and constrict its flow into the main navigation channel. But there was little river traffic in the 6-foot channel. In fact, as Anfinson (1993) states, "By 1918, no through traffic moved between St. Paul and St. Louis." In 1922, the Interstate Commerce Commission declared, "Water competition on the Mississippi River north of St. Louis is no longer recognized as a controlling force but is little more than potential" (cited in Anfinson, 1996).

Boosters of navigation fought strongly to restore commercial shipping on the Mississippi River. Despite a "thirty-year decline in traffic on the upper river" (Anfinson, 1993), in 1930 Congress included a 9-foot-channel project in the 1930 Rivers and Harbors Act. Under this project, the Corps of Engineers constructed Locks and Dams 3 (Red Wing, Minnesota) through 26 (Alton, Illinois), joining completed structures at Keokuk, Iowa (1913), St. Paul, Minnesota (1917), and Hastings, Minnesota (1930).

The 9-foot channel-project represented a turning point in Upper Mississippi River history. Previous efforts at deepening the river's channel focused on narrowing the channel through dredging, wing dams, and closing dams (which extend from one shore to an island, or between islands, to divert flows from backwaters and side channels into the main channel). The locks and dams constructed to help create the 9-foot navigation channel, in contrast, created a series of navigation pools, significantly changing river commerce and ecology (Anfinson, 1993):

> With the nine-foot channel project, the Corps initiated a completely new approach to navigation improvement. The locks and dams would widen and deepen the river, slowing its pace. In doing so, the Corps would create a more reliable navigation channel and enable shippers to match or exceed the economies of scale enjoyed by railroads. The wider, deeper, and slower-moving channel would affect the river's ecosystems in ways far different from channel constriction.

In 1940, the Corps completed the 9-foot channel project. Twenty-six locks and dams now crossed the Upper Mississippi between Minneapolis and Alton. Lower and Upper St. Anthony Falls locks and dams were completed in 1956 and 1963, respectively, and Lock and Dam 27 (Granite City, Illinois) was completed in 1964, bringing the total to 29.

Commercial navigation on the Upper Mississippi grew substantially following completion of the 9-foot project, eventually increasing to 126 million tons per year by 1995 (USACE, 1995). But the navigation-enhancement projects on the Upper Mississippi River, as well as other river training activities and barge traffic, have had negative ecological side effects. Congress has often been at the center of discussions regarding the appropriate balance between navigation and river training activities, and environmental conservation. An example of congressional efforts to accommodate both navigation and environmental interests is in the 1986 Water Resources Development Act (WRDA), in which Congress declared the Upper Mississippi River system "a nationally significant ecosystem and a nationally significant commercial navigation system" (WRDA 1986; Public Law 99-662). As part of this act, the Environmental Management Program (EMP) was established. The EMP oversees long-term ecological monitoring on the Upper Mississippi River, and it plans and constructs fish and wildlife habitat restoration projects. The current federal agencies with responsibilities in the EMP are the Corps (planning, design, construction, and monitoring of habitat restoration projects), the U.S. Fish and Wildlife Service (lead coordination responsibility and other duties), and the U.S. Geological Survey (long-term ecosystem monitoring). In addition to the EMP, Congress also authorized construction of a second lock at Lock and Dam 26 at Alton, Illinois, as part of the 1986 WRDA.

NAVIGATION SYSTEM FEASIBILITY STUDY OF THE WATERWAY SYSTEM

Initial investigations regarding potential expansion of waterway capacity on the Upper Mississippi River and Illinois Waterway were conducted in the late 1980s. Corps of Engineers' planning studies are conducted by the Corps' district offices in two phases: reconnaissance and feasibility. Reconnaissance studies are conducted to determine if there is federal interest in a water resources problem or opportunity. Reconnaissance studies today are to be conducted in no more than one year, are to cost no more than $100,000, and are fully funded by the federal government. A reconnaissance study ends with a recommendation to either halt the planning efforts or to proceed to the feasibility study stage.

Feasibility studies are cost-shared with a local sponsor and are conducted to formulate alternative approaches (e.g., structural, nonstructural) and alternative design characteristics (e.g, different materials that could be used for a flood damage reduction levee, or different levee heights). Among these alternatives, the Corps is mandated to include an economically optimal alternative, the National Economic Development (NED) plan. Once a range of alternatives is identified, the Corps generally convenes an Alternative Formulation Briefing (AFB) for study sponsors, interested groups, and the public. When the study sponsor and the Corps agree on a final plan, the feasibility study is signed by the Corps' division office engineer, signifying that the Corps' field-level office approves of the project. The feasibility study is then submitted to Corps' Headquarters (HQ) in Washington, D.C. for review. The Washington-level review is the first level of formal study review beyond the district office.

The review at Corps Headquarters is conducted by Corps engineering and technical staff. If the feasibility study is approved by Corps Headquarters, the study is forwarded to the Corps' Chief of Engineers for final approval. The approval from the Chief of Engineers results in a "Chief's Report," which is forwarded to the Office of Management and Budget for inclusion in the president's budget (for more detailed discussion of the Corps's water resources planning procedures, see NRC, 1999a).

In May 1988, the Corps' Rock Island district office developed an initial appraisal regarding possible expansions on the waterway. In August 1989, the Rock Island district office completed a Plan of Study (POS) for the UMR–IWW navigation studies. The Corps' Plan of Study (USACE, 1991) recommended two separate navigation reconnaissance studies for investigating potential navigation improvements. Following this recommendation, separate reconnaissance studies were conducted for the Illinois Waterway and the Upper Mississippi River. The three-volume reconnaissance report for the Illinois Waterway was completed in October 1990, and the two-volume Upper Mississippi River reconnaissance report was completed in June 1991. Both reports recommended more detailed systematic analysis of environmental, engineering, and economic issues so that a system wide study of the navigation system could be developed (USACE, 2000a).

In October 1991, Corps of Engineers Headquarters directed that the two studies be combined into one feasibility study providing a systems approach to addressing navigation issues. Staff members from the (then) North Central and Lower Mississippi Valley Divisions of the Corps, and from its St. Paul, Rock Island, and St. Louis district offices, developed a strategy for conducting the feasibility study. In 1997, the Corps reorganized its division offices, and these three district offices became part of the Corps' new Mississippi Valley Division (along with Corps district offices in Memphis, New Orleans, and Vicksburg). Staff from the New Orleans district office of the Corps also joined the study group.

The Corps' navigation feasibility study of the Upper Mississippi River and Illinois Waterway system was initiated in April 1993. This feasibility study followed a Corps reconnaissance study that indicated the potential need for immediate improvements at several locks and dams and future improvements at others. The Corps stated that "the principal problem being addressed is the potential for significant traffic delays on the system within the 50-year planning horizon, delays that will result in economic losses to the Nation" (USACE, 2000a).

The study was originally envisioned to be a 6-year effort, but because of its complexity and because of comments from the public and coordinating agencies, some modifications were necessary. For example, in early 1998, it was recognized that some of the environmental, economic, and engineering studies were taking longer to complete and review than originally anticipated.

The Corps divided research and planning duties within the study into five groups: economics, engineering, environmental/historic properties, project management/plan formulation, and public involvement. In addition to holding public meetings about the navigation study, the Corps worked with other stakeholders and experts from other federal and state agencies. These groups included a Governors' Liaison Committee—which consisted of governor-appointed representatives from the five Upper Mississippi River basin states (Illinois, Iowa, Minnesota, Missouri, and Wisconsin) and the commander of the Corps' Mississippi Valley Division—and the Navigation Environmental Coordination Committee (NECC), which consisted of representatives from the five basin state-level natural resources and conservation agencies.

The Corps issued a draft feasibility study report in July 2000 (USACE, 2000a) and was scheduled to release its final report in the second half of 2001. This National Research Council study and report thus reviews and assesses the Corps' draft feasibility study. As of late 2000, the cost of the draft navigation feasibility study was $55.6 million, with the largest percentage (43 percent) of this cost being associated with the environmental analysis (Figure 1.2).

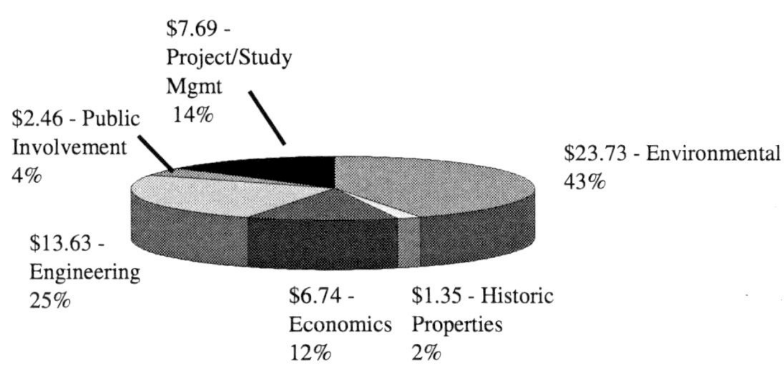

FIGURE 1.2 Feasibility Study Cost Estimate ($ in millions and percent of total).
SOURCE: USACE, 2000a.

CHARGE TO THE COMMITTEE

In early 2000, the DOD requested the National Academies to review the Corps' Upper Mississippi River–Illinois Waterway navigation system feasibility study. Given the publicity surrounding this study, it was requested that the study be completed relatively quickly. A joint committee of the National Academies' Transportation Research Board (TRB) and Water Science and Technology Board (WSTB) was convened to carry out the study. The charge to the committee was as follows:

> This study will focus on the U.S. Army Corps of Engineers' economic analysis regarding proposed improvements, including economic assumptions, methods, and forecasts regarding barge transportation demand on the Upper Mississippi River-Illinois Waterway. The Corps must also consider larger water resources project planning issues, such as formal U.S. federal water resource planning guidelines, possible environmental impacts, and the costs of navigation improvements. Thus, while the committee will focus on the Corps' economic analysis, they will also comment upon the extent to which these larger issues are being appropriately considered in the navigation system feasibility study.

Because the Corps has not yet completed its navigation feasibility study, the committee could not review and comment on the Corps' final navigation feasibility study; the committee, however, reviewed a draft of that study (USACE, 2000a). The committee concluded, and the Department of Defense agreed, that an interim report that reviewed and commented on the Corps' models and methods would be useful. Based on its review of the Corps' methods in the draft report, the committee aims to provide helpful advice to the Corps for completion of the navigation system feasibility study.

2

The Public Interest in the Upper Mississippi River

Investment in and management of an inland waterway is affected by the fact that almost all investment in navigation enhancement and river-training facilities is public and almost all use of the waterway is private. No federal agency would want to assume direct control over the multiple uses of inland waterways. Privatizing these facilities and services is an even less attractive option. A company that controlled commercial navigation would find itself making decisions that affected not only navigation, but also municipal water supply, recreation, irrigation, flood damage reduction, and environmental quality. Privatization would not work well unless the controlling firms faced the proper incentives regarding each possible use of the waterway. There are also disagreements over the goals to be achieved in managing a waterway; a proposal to turn over waterway operation and maintenance to a private firm would be met with intense opposition from almost all constituents. Thus, tensions of public ownership and private use are inherent in U.S. inland waterway systems.

Under present management practices, shippers and towboat operators are mainly motivated to consider their own costs while neglecting the government's costs. For example, a current controversy is whether "industry self-help" should be used to reduce congestion by speeding the locking process. Industry self-help imposes costs on towboat operators, but costs the government nothing. In contrast, extending locks shifts the costs of reducing congestion from towboat operators to the government (a portion of the project costs are paid by the fuel tax that goes into the Inland Waterways Trust Fund[1]).

The federal government, through the Corps, is responsible both for investing in navigable waterways and for managing their operations and for minimizing or mitigating environmental side effects. The implicit goal of inland waterway management has been to provide low cost freight transportation to all users. This goal has been manifested as providing equal access to all on a "first-come, first-served" basis, and expanding the system when congestion becomes a

[1] The Inland Waterways Trust Fund was authorized by two separate acts of Congress (1978 and 1986) and imposes a waterway fuel tax on users of many U.S. inland waterways. The tax has increased over time, starting at a rate of 4 cents per gallon in 1980, increasing to a rate of 20 cents per gallon after 1994. As of the year 2000, the total in the fund stood at $388 million. The fund expenditures are transferred to the Corps for disbursement to contractors or to reimburse Corps civil works for in-house costs.

problem. Shippers have been reluctant to seek innovative, nonstructural ways to reduce congestion; reductions in congestion would reduce river traffic (at least at peak times), thereby reducing the benefits of lock extensions. Unfortunately, the "first-come, first-served" rule is an inefficient way to manage river traffic, as it results in higher systems costs (costs to all shippers) than if there were a system explicitly designed to reduce congestion. Better management of waterway traffic should result in improved service and lower total shipping costs—with benefits to most waterway shippers. In particular, farmers would benefit from lower shipping costs.

Large amounts of waterway traffic, and the hydrodynamic changes caused by the series of navigation pools, have effects on aquatic habitat and species. The many federal, state, and local environmental conservation laws, such as the Clean Water Act and the Endangered Species Act, reflect public concern for protecting the river and its ecosystems. In addition, environmental protection provides tangible benefits from tourism, recreation, and the production of food and fiber. Estimates of the annual revenue generated by tourism and recreation in the Upper Mississippi range from $1.2 billion (Carlson et al., 1995) to $6.6 billion (cited in UMRCC, 2000). The vast river–floodplain ecosystem of the Upper Mississippi River basin also provides a range of ecosystem services, including drinking water, food (fishes and waterfowl), groundwater recharge, purification of polluted waters, and flood retention. The Upper Mississippi River ecosystem is a storehouse of biodiversity, which produces social benefits today (e.g., food and fiber), and may produce additional benefits in the future (e.g., medicines).

On the other hand, intensive use of the waterway has negative effects on river ecology and, in turn, on these various social values and goods. The construction and subsequent operation of the dams and navigation pools on the UMR–IWW has also resulted in a range of environmental effects. Given these external costs, the multiple uses of a river and waterway system must be considered explicitly when deciding how much traffic should be permitted on the waterway and whether locks should be extended in order to accommodate more traffic. The public interest would be best served by river traffic management practices that are environmentally sustainable; that is, strategies that promote both a better flow of river traffic and the maintenance of ecosystem habitat and processes.

The Corps has viewed its responsibility as providing adequate capacity to serve all waterway users; for example, deciding on the proper investments and determining the best time to make those investments. These decisions are guided by federal laws and congressional guidance (a fuller discussion of which is provided in Chapter 3). The Corps has given little or no attention to allocating the waterway among all those who wish to use the locks when there is congestion.

The best solution to the problem of waterway congestion would be to simultaneously optimize access to the locks and to determine when public investment to extend the locks might be warranted. Instead, access to the locks is determined on the basis of delays caused by having to wait for others to clear the lock. Long waits to transit the locks indicate either that the locks should be extended or that current demand for the locks is being managed poorly. The public interest requires that the relevant government agency have responsibility for both managing the traffic and investing in lock extensions. Approaches for traffic management include nonstructural options such as tradable permits, congestion tolls, scheduling traffic, and charging for the time taken to transit a lock. This management problem interacts with a larger, more contentious one. Midwest grain could be moved on several routes for export. In addition to going by barge to New Orleans, grain could be shipped by rail to St. Louis and then by barge to New Orleans. It could also be shipped by rail to the Great Lakes or by rail to the Columbia River, then by barge or by rail to West Coast ports such as Portland. Thus, in addition to farmers, towboat operators, and recreational users, railroads, and ports (such as New Orleans, St. Louis, and Portland) also have a large stake in the issue of lock extensions and grain transportation on the UMR–IWW.

Demand for waterway transportation is a derived demand that depends on demand for grain in other nations and the cost of shipping grain to the customer. Iowa grain competes with Illinois grain, as well as with Argentinian, Brazilian, and Canadian grain. Shipping costs are an important determinant of which grain is cheapest in Yokohama or Shanghai. Congestion at Mississippi River locks could result in Brazil getting the sale and Iowa land being used for other crops. The combination of fertile land in the Upper Midwest and an extensive water transport system has conveyed a large advantage to American grain.

Fundamentally, Congress directs the Corps to conduct the wrong analysis in assessing the benefits and costs of lock extensions under the current approach to waterway traffic management. Improving traffic management is a more immediate, cost-effective, and environmentally sustainable way to handle congestion than through traditional capacity (supply side) expansion. Because traffic is managed by waiting time and service rules that were not designed to internalize congestion costs, lock demand is artificially high. If the benefits of lock extensions are based on waterway traffic levels without any type of traffic management system, the analysis will thus overstate the social benefits of extensions and could lead to lock extensions where none are justified. If traffic on the waterway was properly managed, the economic justification for some lock extensions would disappear. Congress has the authority to determine public policies for inland waterways and has the responsibility to make such determinations. If Congress believes that the public interest is served by subsidizing navigation on these waterways, it is within Congress' authority to subsidize navigation. Past decisions by Congress and the Corps imply that both groups believe that commercial traffic on inland waterways is of great value to the public. Benefits are thus estimated on the basis of serving demand, where the level of demand is determined by private costs, rather than by the higher, collective costs of all of the system's users. We note this implicit assumption and the bias in the resulting analysis. Having done that, we now proceed to advise the Corps on how to conduct a benefits analysis under their assumptions.

Allocating Use of a Waterway

In Garrett Hardin's classic commons problem (Hardin, 1968), it is assumed that herdsmen are motivated by their individual benefits and costs. Because the benefits to a herdsman of grazing an additional animal on a communal pasture (the commons) outweigh his costs, every herdsman adds animals to the common, eventually resulting in overuse and gross mismanagement of the public good.

On the UMR–IWW, shippers, towboat operators, and others who desire to use the waterway for navigation, recreation, and other purposes, likewise view their benefits of using the waterway as exceeding their costs of using the waterway. As a result, the Upper Mississippi River–Illinois Waterway is intensively used. This results in increased congestion and environmental degradation. Several ecological indicators on the Upper Mississippi River—such as the viability of native populations and their habitats, the ability of the ecosystem to recover from disturbances, and ecosystem sustainability—are in decline (USGS, 1999). The causes of these effects are multiple and difficult to precisely ascribe. Commercial navigation is but one of several factors that have affected the river ecosystem. The Corps projects additional congestion in the future, based on projections of increased grain exports from this region.

The Upper Mississippi River–Illinois Waterway serves a diverse clientele who seek different services. Furthermore, some uses and users impinge negatively on others in "one-way" types of externalities. For example, barges on the river have significant and damaging effects on fish; but fish do not negatively affect the barge industry. Striking the proper balance between the multiple uses and users and thereby protecting the public interest requires denying some potential users access when they want it. Rationing access could be done by regulation: for example, commercial traffic could be given priority over recreational boating, or tows could be forced to sign up for a schedule long in advance. Rationing could also be done by pricing, since some of these users would be willing to pay a great deal to use these services, while others would be willing to pay almost nothing. The methods of allocating access differ with respect to their efficiency and equity. The point is that allocating access is better than overusing a public resource.

If access is allocated by charging for it, there is a natural relationship between a congestion toll and the decision to expand capacity. A congestion toll provides a monetary value of the user's willingness to pay for use of the lock and is a clear indication of the public benefits of expanding capacity. In contrast, when there is no toll for use, users' willingness to pay for use of the waterway is not as clearly expressed.

Investment decisions for the UMR–IWW system are complicated by (1) multiple values, uses, and users of these waterways, (2) contradictions between public ownership and private use, and (3) the lack of a system for managing congestion, which leads to market failure. The first two problems are inherent in the system. The third problem requires a solution before intelligent decisions can be made about investments in increasing lock capacity.

3

U.S. Federal and Corps of Engineers Water Resources Planning Guidelines

Guidance for the Corps of Engineers' civil works program is provided by the federal *Principles and Guidelines for Water and Related Land Resources Implementation Studies*, developed by the U.S. Water Resources Council (WRC). This document is also referred to as the *Principles and Guidelines*, or the *P&G*. The Corps has incorporated these planning guidelines in its Planning Guidance Notebook, Engineer Regulation (ER) 1105-2-100 (USACE, 2000b). The Corps also publishes many other engineering regulations, several of which affect its inland waterways programs, either as supplements to the *P&G* or that cover processes beyond the scope of the *P&G*.

In the committee's examination of the Corps' draft Upper Mississippi-Illinois Waterway System navigation feasibility study, it is appropriate to begin with a review and critique of the *P&G*, a document that has remained unchanged since 1983. Particular attention is given to the federal objective, environmental quality objectives, advanced analytical methods and datasets, and the treatment of uncertainty in formulating and evaluating project alternatives.

ESTABLISHING THE *PRINCIPLES AND GUIDELINES*

Efforts to develop criteria to guide the formulation and evaluation of waterways projects extend as far back as 1808, when U.S. Secretary of the Treasury Albert Gallatin published a report identifying opportunities to develop a waterway-based transportation system. The Reclamation Act of 1902 contained guidance for water resources project evaluation, and the Corps used a form of benefit–cost analysis to evaluate basinwide plans prepared in the 1920s and 1930s pursuant to the Rivers and Harbors Act of 1927. In addition to economic objectives, much of the major water legislation of the first four decades of the 20th century also had social objectives.

Congressional policy was established by the 1936 Flood Control Act, which stated that: "the federal government should improve or participate in the improvement of navigable waters or their tributaries, including watersheds thereof, for flood control purposes if the benefits to whomsoever they accrue are in excess of the estimated costs, and if the lives and social security of people are otherwise adversely affected." This is the original legislation justifying federal water resources projects if benefits exceed costs. However, Congress stopped short of directing the administration to develop methods for implementing this policy. Following the demise of the National Resources Planning Board, President Roosevelt established the Federal Interagency River Basin Committee (FIARBC) in 1943. One of its tasks was the development of *Proposed Practices for Economic Analysis of River Basin Projects*, originally issued in 1950 and revised in 1958. The Kennedy Administration, acting on a 1960 recommendation of the Senate Select Committee on Water Resources, further revised guidance for water project planning and evaluation by preparing *Policies, Standards and Procedures in the Formulation, Evaluation, and Review of Plans for Use and Development of Water and Related Land Resources,* published in 1962 as Senate Document 97, 87th Congress, Second Session.

The Water Resources Planning Act of 1965 implemented a number of recommendations of the Senate Select Committee, including establishment of the Water Resources Council (WRC), a Cabinet-level council of the secretaries of all federal departments involved in water management. One of the charges to the WRC was to promulgate principles and standards for water projects. That was done when the WRC published the *Principles and Standards for Planning Water and Related Land Resources* September 10, 1973, often referred to as the *Principles and Standards*, or the *P&S* (US WRC, 1973).

The *Principles and Standards* departed from prior policy by establishing a multiple-objective planning framework. Previous policy directed federal agencies to evaluate all inputs and outputs of projects in economic terms to the extent possible and to describe nonmonetary effects as "intangible" benefits and costs. With the publication of the *P&S* as federal regulations, the WRC established national economic development (NED) and environmental quality (EQ) accounts as co-equal objectives of federal water projects. Federal water management agencies also were directed by the *P&S* to evaluate beneficial and adverse effects of projects on regional economic development (RED) and other social effects (OSE).

President Carter's administration further modified the *P&S* to require greater attention to environmental effects of water projects. By making water projects more difficult to justify, President Carter alienated a number of western interests. When the Reagan Administration took office in 1981, the Water Resources Council and its *P&S* had become prime political targets. The WRC's budget was zero-funded by the Reagan administration and the WRC convened in 1983 only long enough to repeal the *Principles and Standards*, replacing it with the *Principles and Guidelines* (US WRC, 1983).

There was more to that change than simply a modification of the title. A major change was that the *Principles and Standards* was reduced to the level of guidance, no longer having the legal standing and enforceability of federal regulations.

The Federal Objective

Another modification that appears to represent a significant change from the *Principles and Standards* to the *Principles and Guidelines* was the shift in the status of the environmental quality (EQ) objective. In the *P&S,* NED and EQ were given equal weighting as federal objectives. By contrast, the *P&G* specified a single objective that stated that water resources projects are "to contribute to NED consistent with protecting the Nation's environment, pursuant to national environmental statues, applicable executive orders, and other Federal planning requirements" (US WRC, 1983). One interpretation of this change is that while the *P&S* treated the environmental quality account with equal importance to the NED account, under the *P&G*, environmental quality was relegated to a constraint on the alternative that maximizes contributions to the NED objective.

Under the *P&S*, federal agencies were required to evaluate proposed projects using four separate accounts: (1) National Economic Development, (2) Environmental Quality, (3) Regional Economic Development, and (4) Other Social Effects. The *Principles and Guidelines* directed the Corps to include among alternative plans, one that maximizes NED consistent with environmental constraints, and others that reduce NED benefits in order to address other federal, state, local, and international concerns. Of the four accounts established by *Principles and Standards*, only the NED alternative is required by *Principles and Guidelines.* One or more of the other three accounts are necessary only if they "will have a material bearing on decisionmaking" or if required by law.

These changes may be more apparent than real for at least two reasons. First, it was never clear to water resources project planners how NED and EQ were to be made coequal in planning under *P&S*. Contributions to the two objectives were measured by very different metrics, and no set of weights that makes x number of habitat units equal to y million $ has yet been established. These types of decisions are thus judgments within the discretion of the Corps, its constituents, the public, and the Congress. Second, the *Principles and Guidelines* did not limit planning strictly to the maximization of NED. In the section on "Alternative Plans," the *P&G* state (US WRC, 1983):

> In addition to a plan which reasonably maximizes contributions to NED, other plans may be formulated which reduce net NED benefits in order to further address other Federal, State, local and national concerns not fully addressed by the NED plan. These additional plans should be formulated in order to allow the decisionmaker the opportunity to judge whether these beneficial effects outweigh the corresponding NED losses.

This guidance would appear to provide the Corps considerable flexibility in formulating alternatives, not only to promote national economic development but to address environmental issues, as well. Although all federal agencies that use the *P&G* (in addition to the Corps, these agencies are the Bureau of Reclamation, the Natural Resources Conservation Service, and the Tennessee Valley Authority) are directed to formulate an alternative that reasonably maximizes NED, there is no apparent limit on the extent to which the Corps can go in formulating and evaluating other alternatives to improve environmental quality. How the Corps of Engineers chooses to interpret and apply this guidance to particular projects can have a profound effect on how alternatives are formulated and evaluated, but the *P&G* is not overly restrictive.

National Economic Development

The NED account reflects benefits and costs of a plan, namely increases in the economic value of goods and services and the opportunity and other costs of resources consumed by the plan. A national accounting viewpoint is to be considered. Any change in one region resulting from an interregional transfer is not to be counted.

The *Principles and Guidelines* contains more specific guidance for estimating beneficial effects of particular kinds of goods and services (inland navigation being the one most relevant to this study). Four categories of navigation benefits are described:

1. cost-reduction benefit—the gain in economic efficiency when traffic uses a waterway with and without the plan with the same origins and destinations and the same mode,
2. shift-of-mode-benefit—the reduction in cost when traffic would move from the same origins and destinations, but with the plan, it would move by waterway instead of the more costly alternative mode,
3. shift-of-origin destination benefit—the reduction in total cost of getting a commodity to its place of use when its origin is shifted from its location without the plan to a new location with the plan, or the change in net revenue to the producer if the destination with the plan is different from the destination without the plan, and
4. new movement benefit—the increase in producer and consumer surplus associated with increased transport of a commodity due solely to reduced transportation cost.

Water Resources Project Planning

The planning process outlined in the *Principles and Guidelines* consists of six major steps:

Step 1 - Identify problems and opportunities. The first step includes an "early and open" scoping process to identify the types and extent of issues relevant to the study. It is intended to complement the scoping process required under the National Environmental Policy Act (NEPA) process. It should identify the geographic areas likely to be affected by alternative plans.

Step 2 - Inventory and forecast conditions. Economic, environmental, and other factors affected by the plan must be forecast over the planning horizon with and without the plan. An inventory of existing conditions should be used as the baseline, and future conditions should reflect national and regional projections of income, output, and other relevant information published by federal agencies (insofar as the agency forecasts are not intended to be projections or if they do not track the observed data since the forecasts were made, the Corps should develop their own forecasts, using methods recommended in Chapter 5 of this report).

Step 3 – Formulate alternative plans. Under Section 1.6.2 of the *Principles and Guidelines,* federal agencies are provided guidance concerning the kinds of alternatives that are to be formulated. Agencies are required to formulate a plan that reasonably maximizes contributions to NED, but other plans may be formulated that reduce NED benefits to address other federal, state, local, and international concerns not captured under the NED objective. The purpose of alternative plans is to aid decision-makers in their judgment as to whether the benefits added by the alternatives exceed reduction in benefits from the NED plan.

That section of the *Principles and Guidelines* is even more specific on how alternatives should be formulated. It states that in formulating alternative plans, an effort is to be made to include only increments that provide NED benefits after accounting for appropriate mitigation costs. Appropriate mitigation of adverse environmental effects, as required by law, is to be included in all alternative plans. Increments that do not provide net NED benefits may be included, except in the NED plan, if they are cost-effective measures for addressing specific concerns.

Considerable guidance is provided by the *Principles and Guidelines* that affect the content of alternative plans and how certain elements of alternatives are to be treated. Two sets of assumptions are listed in the *P&G*: a without-project condition, and a with-project condition, as listed below (US WRC, 1983, Sec. 2.6.3):

Without-project condition:

- Assume that all reasonably expected nonstructural practices within the discretion of the agency, including helper boats and lock operating policies, are implemented at the appropriate time. Substantial analysis is required to determine the best combination of nonstructural measures to ensure the most effective use of an existing waterway system over time. This analysis should be documented in project reports to assure the reviewer that the best use of existing facilities will be made in the without-project-condition and that the benefits of alternative with-project conditions are correctly stated. The criteria for the best utilization of the system are overall public interest concerns, including economic efficiency, safety, and environmental impact.
- User charges and/or taxes required by law are part of the without-project condition. Proposed or possible fees, charges, or taxes are not part the without-project condition but should be considered as part of any nonstructural alternatives in the with-project condition.
- The without-project condition assumes that normal operation and maintenance will be performed on the waterway system over the period of the analysis.
- In projecting traffic movements on other modes (railroad, highway, pipeline, or other), the without-project condition normally assumes that the alternative modes have sufficient capacity to move traffic at current rates unless there is specific evidence to the contrary.
- Alternative modes should be analyzed as a basis for identifying the most likely route by which commodities will be transported in the future in the absence of waterway improvement.
- The without-project condition normally assumes that only waterway investments currently in place or under construction are in place over the period of analysis.

With-project condition:

- Management of demand by the use of congestion or lockage fees is a nonstructural alternative, which alone or in combination with structural devices may produce an economic optimum in a congested waterway. Influencing marginal waterway users through a congestion fee can increase the benefits of a waterway. Evaluate alternatives that influence demand on the same basis as supply-increasing (structural) alternatives.
- Additional nonstructural measures not within the current purview of the operating agency may be considered 'supply management' measures. One example is traffic management. These supply-increasing (nonstructural) measures can be used alone or in combination with other structural or nonstructural measures.
- Project alternatives can differ in their timing as well as in their physical characteristics. Consider the optimal timing of projects and of individual project features in project formulation, so as to maximize net benefits over time.
- Consider improvements in alternative transportation modes as part of the without-project condition only.
- A change in the waterway system that is currently authorized but not yet under construction may be included if an appropriate share of its associated costs is included in the costs of the alternative under study and its incremental contribution to benefits is explicitly identified.

Step 4 – Evaluate alternative plans. The *P&G* outlines a 10-step evaluation procedure for estimating NED benefits of inland navigation projects (Sec. 2.6. The procedure is shown in Figure 3.1). Unfortunately, there is only general guidance for forecasting potential waterway traffic and for forecasting modal choice under future costs conditions. The *Principles and Guidelines* identify other problems that may be encountered in applying the procedure. First, differences in system delays between alternative plans are difficult to estimate, requiring state-of-the-art techniques. Second, the procedure in Figure 3.1 assumes that supply and demand curves are independent of each other when in fact they are not. That interdependence is especially important at high levels of lock congestion. Third, uncertainty arises from several factors, and sensitivity analysis is recommended to evaluate the effects of those uncertainties.

Several specific sensitivity analyses are prescribed. One is to assume no growth in tonnage and no change in fleet characteristics after the first 20 years of the projects. Another is to estimate how benefits would be affected if costs were to be fully recovered through user charges.

Step 5 - Compare alternative plans. Specific formats are offered to facilitate the display of findings.

Step 6 - Select recommended plan. A recommended plan is to be selected after "consideration of the various alternative plans, their effects and public comments."

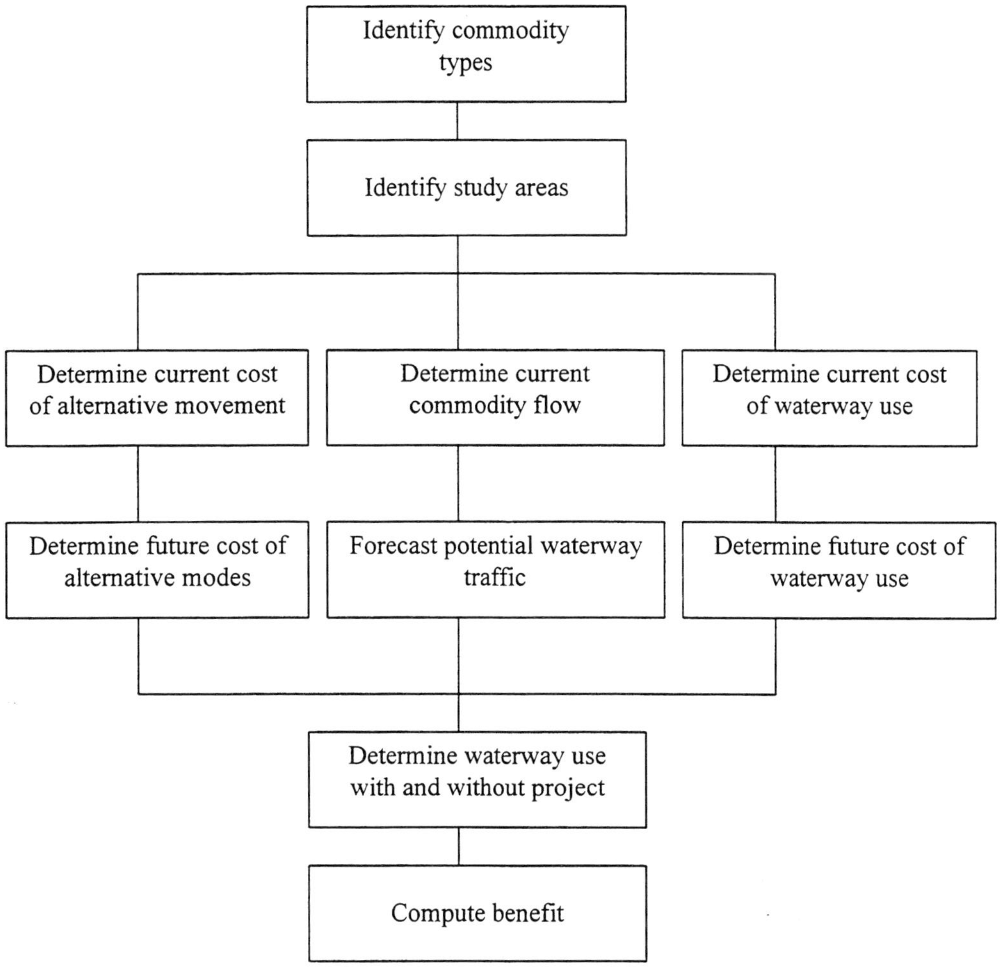

FIGURE 3.1 The 10 Steps for Evaluating Alternative Plans. SOURCE: US WRC, 1983.

CRITIQUE

Although the *Principles and Guidelines* has not been altered since 1983, that document outlines a useful framework for planning inland navigation projects. Despite its general utility, however, the guidance is limited in several ways. First, limitations arise simply because the guidance has not been updated to reflect recent policy changes, such as the inclusion of environmental restoration as a federal purpose. Second, in several places the guidance refers to particular datasets and procedures. Those are now 18 years out-of-date. The *P&G* contains some shortcomings specific to the kind of inland navigation studies that are the subject of this report. Among them are:

- Inadequate guidance on implementing the federal objective on environmental quality, specifically for projects for which environmental restoration is an authorized purpose;
- The failure to provide updated guidance on advanced techniques for forecasting commodity flows, predicting modal choice for shipment of commodities, and forecasting waterway traffic with and without planned improvements; and
- Inadequate guidance on the treatment of uncertainty in project formulation.

Stating Environmental Quality Objectives

The first limitation is inadequate guidance on implementing operational the federal objective on environmental quality in the setting of the UMR–IWW study. Agencies are directed to formulate an NED plan, including the cost of mitigating adverse environmental effects not accounted for in the NED account. Alternatives are to be formulated to address other specific concerns by adding increments to the NED plan. Taken out of context, such an approach could be interpreted as leaving little room for addressing existing environmental problems including, in this instance, those created by the construction and operations of the existing navigation system. However, the *Principles and Guidelines* also provides guidance that plans are to be formulated "which reduce net NED benefits in order to further address other federal, state, local, and international concerns not fully addressed by the NED plan." (US WRC, 1983; 5(b)).

The UMR–IWW navigation feasibility study is authorized by the 1970 Flood Control Act (P.L. 91-611). Section 216 of that act states:

> The Secretary of the Army, acting through the Chief of Engineers, is authorized to review the operation of projects the construction of which has been completed and which were constructed by the Corps of Engineers in the interest of navigation, flood control, water supply, and related purposes, when found advisable due to significantly changed physical or economic conditions, and to report thereon to Congress with recommendations on the advisability of modifying the structures or their operation, and for improving the quality of the environment in the overall public interest.

The draft navigation feasibility study appears to have interpreted that authorization narrowly. For example, the purpose in the draft feasibility study fails to mention possible operational changes and immediately zeroes in on capital investments. The draft study also fails to embrace the broader authority to recommend improvements to the quality of the environment, and fails to go beyond simply identifing impacts associated with new facilities and increased traffic that are to be mitigated. In short, the navigation feasibility study focuses narrowly on the federal objective as stated in *P&G,* which is to contribute to national economic development, consistent with protecting the nation's environment. The feasibility study does not incorporate operational and environmental alternatives consistent with the project-specific directive from Congress, despite the fact that the Corps appears to have ample latitude to explore alternatives aimed at enhancing environmental resources according to (1) the 1970 Flood Control Act, (2) the *Principles and Guidelines*, which allows for the formulation of plans that reduce new NED benefits in order to further address other concerns not fully addressed by the NED plan, and Corps planning guidance that allows for the consideration of a National Ecosystem Restoration objective (USACE, 2000a; that guidance, however, was established after the feasibility study was initiated). Even though legislation and planning guidance such as the *P&G* provides room for the Corps to consider environmental problems and opportunities as part of the navigation feasibility study, alternatives to improve environmental quality were not adequately considered by the Corps. The reasons for this are not clear to this committee. The Secretary of the Army should ensure that the environmental consequences of proposed construction and operating practices be analyzed along with the NED. Environmental improvements—not just the mitigation of incremental environmental damages—should also be examined.

Advanced Analytical Methods and Datasets

A second limitation of the *Principles and Guidelines* is the failure to provide updated guidance on advanced techniques for forecasting commodity flows, predicting modal choice for shipment of commodities, and forecasting waterway traffic levels. While the *P&G* does not prevent the Corps or other federal agencies from using advanced analytical techniques, it fails to offer guidance as to which methods are most appropriate. The UMR–IWW is one of the world's larger and more intensively used navigation systems. It serves a complex agricultural and industrial region where substantial inter-modal competition exists. Exports from the region are being sold in increasingly globally competitive markets. Estimated costs of proposed lock extensions are in excess of one billion dollars. Use of the best available analytical techniques is justified for a system as large and complex as the UMR–IWW.

As Chapters 4 and 5 will discuss, a regional econometric model for forecasting future market conditions, with linkages to key policy decisions at the national and international level, would provide greater confidence in projections of future demand for transport. The credibility of traffic forecasts would be enhanced by using models that consider explicitly how the cargo could move by alternative modes and that examine geographic areas smaller than a state in order to recognize the diversity in yield, market opportunities, and shipping costs for local areas. Without such methods, developed with the benefit of an open, interdisciplinary, peer-reviewed process, agency staff are left to their own devices to develop ad hoc methods within project budgetary and time constraints.

Uncertainty in Formulation and Evaluation of Alternatives

A third shortcoming of the *P&G* is the inadequate guidance on the treatment of uncertainty in the formulation of alternatives. Although the *Principles and Guidelines* directs agencies to give consideration to uncertainty, attention is given primarily to uncertainties in parametersused in models for predicting future conditions and for the detailed design of facilities. Chapter 5 of this report describes approaches for addressing uncertainty.

Revising the *Principles and Guidelines*

The *Principles and Guidelines* should be reviewed and updated in order to reflect changes in social values since the early 1980s, as well as advances in analytical techniques and technologies. This recommendation is consistent with the recommendation of a 1999 National Research Council committee charged to, among other things, review the state of the *P&G*. That committee recommended that the federal *Principles and Guidelines* "be thoroughly reviewed and modified to incorporate contemporary analytical techniques and changes in public values and federal agency programs" (NRC, 1999a; see also NRC, 2000). The 1994 federal Interagency Floodplain Management Review Committee (IFMRC) report, also known as the Galloway Report (as the review committee was headed by Dr. Gerald E. Galloway, Jr.), contained a similar recommendation: "*Principles and Guidelines* should be revised to accommodate the new objectives and to ensure full consideration of nonstructural alternatives" (IFMRC, 1994). In revising the *P&G*, special attention should be given to: (1) guidance for incorporating concerns about environmental quality into plans for inland navigation systems, regardless of how the EQ objective is stated, (2) more definitive guidance on techniques for forecasting demands for waterway services over extended time horizons, and (3) more appropriate methods for treating uncertainty in large projects that are subject to high levels of uncertainty in future demand.

The *P&G* has provided useful planning guidance since it was approved in 1983, with much of that guidance still relevant and useful today. After 18 years of application, however, the *P&G* would be improved with a thorough review and the incorporation of modern water resources planning approaches and techniques.

4

The Navigation Feasibility Study

This chapter examines the economic, environmental, and engineering aspects of the U.S. Army Corps of Engineers' draft feasibility study of the Upper Mississippi River–Illinois Waterway system, based on the most recent and complete information that was available to the committee. As mentioned, the committee reviewed and commented on a draft of the feasibility study, as well as on other important, supporting studies the committee was directed to by the Corps. This report thus may not reflect the Corps' final navigation feasibility study.

As described in its charge, the committee focused its review on the Corps' economic analyses, as well as other relevant water resources planning considerations. Thus, most members of this committee were economists, and the committee devoted most of its efforts toward assessing the economics portions of the Corps' study. The Corps' draft feasibility study contained three main topical areas: economics, engineering, and environment (see Figure 1.2). The quality and detail of the analyses conducted within each of these topics varied. For example, the structure and logic of the economics sections were more coherent and more sophisticated than the environmental study components. This report's treatment of economics, engineering, and environmental issues in the feasibility study therefore reflects the committee's charge, the disciplinary expertise of committee members, and the varying levels of analysis between different sections of the Corps' draft report.

PROJECT SCOPING, ALTERNATIVES, AND INTEGRATION

Plan Formulation

The *Principles and Guidelines* require that combinations of measures be considered in order to identify the environmentally acceptable water resources development plan that produces the greatest net benefits to the national economy. Identifying this National Economic Development (NED) plan involves assessing the economic and environmental impacts of a range of different capacity-enhancing and demand-management measures, alone and in combination with other measures. It is important to recognize that the *Principles and Guidelines* also provide that, "Other plans which reduce new NED benefits in order to further address other Federal, State, local, and international concerns not fully addressed by the NED plan should also be formulated" (US WRC, 1983).

As described in Chapter 3, the 1970 Flood Control Act authorizes the Corps to consider navigation, flood control, and water supply project operation alternatives, as well as alternatives focused on improving environmental quality. However, the study purpose and scope in the UMR–IWW navigation draft feasibility study was framed in comparatively narrow terms (USACE, 2000b):

> The Navigation Study is a feasibility study addressing navigation improvement planning for the UMR–IWW System for the years 2000–2050. This study assesses the need for navigation improvements at 29 lock and dam sites (35 locks) on the UMR and 8 locks on the IWW and the impacts of providing these improvements. More specifically, the principal problem being addressed is the potential for significant traffic delays on the system within the 50-year planning horizon, delays that will result in economic losses to the Nation.
>
> The study was conducted to determine whether navigation improvements were justified and, if so, the appropriate navigation improvements, sites, and sequencing for the 50-year planning horizon. The feasibility study effort also included the preparation of a system Environmental Impact Statement (EIS) (USACE, 2000a).

This statement of study scope does not include consideration of alternatives aimed at enhancing environmental quality. Instead, it focuses solely on capital investments, with little attention devoted to facility operational alternatives. Moreover, environmental considerations are relegated to the status of constraints, rather than considered as planning objectives. The Corps thus ignored guidelines that allow them to address a wider array of environmental issues.

The Corps has recently addressed the extent to which environmental improvement may be used as a planning objective, toward which alternatives may be formulated and evaluated. Chapter 2 of the Corps' *Civil Works Planning Guidance Notebook* defines a National Ecosystem Restoration (NER) objective, defining NER outputs as "increases in the net quantity and/or quality of desired ecosystem resources" (USACE, 2000a). This planning guidance was released well after the Corps began its feasibility study. Thus, rather than holding the Corps and the draft feasibility study to these recently-released standards, the committee suggests that this and other recent guidance be followed in future revisions to the feasibility study.

Another provision of this internal Corps planning guidance is the following:

> Measurement of NER is based on changes in ecological resource quality as a function of improvement in habitat quality and/or quantity and expresses quantitatively in physical units or indexes (but not monetary units). These net changes are measured in the planning area and in the rest of the Nation. *Single purpose ecosystem restoration plans shall be formulated and evaluated in terms of their net contributions to increases in ecosystem value (NER outputs), expressed in*

> *non-monetary terms*. Multipurpose plans that include ecosystem restoration shall contribute to both NED outputs and NER outputs (emphasis added).

In addition to these higher-level considerations of project scope, Chapter 2 of the *Principles and Guidelines* contains additional guidance on scoping and alternatives that does not appear to have been adequately considered in the UMR–IWW study planning process. As Chapter 3 described, the *P&G* provide abundant advice regarding nonstructural alternatives for helping relieve congestion on inland waterway systems. The Corps' draft feasibility study, unfortunately, contains no explanation as to why these types of nonstructural alternatives were ignored (except for helper boats).

The *P&G* offers adequate flexibility for the Corps to investigate a wide range of planning options, as does the Corps' own internal guidance. It appears, however, that only a relatively narrow range of navigation project planning alternatives was considered in the draft feasibility study. As a result, the Secretary of the Army should ensure that the environmental consequences of proposed construction and operating practices be analyzed along with the National Economic Development account. In the committee's judgment, environmental issues will continue to grow in importance. By restricting its study to remediating the incremental changes caused by lock extensions, the Corps is creating future problems for the project. The Corps should address a wider range of environmental issues in the UMR–IWW feasibility study, rather than limiting the study to mitigating the environmental damage caused by lock extensions and the resulting incremental increases in barge traffic.

Given the relatively narrow range of alternatives considered for addressing waterway congestion on the UMR–IWW, Congress should instruct the Corps to first consider nonstructural options for improving traffic management as the baseline condition for the NED alternative. Lock extensions should not be considered until nonstructural measures for waterway traffic management have been examined thoroughly. A careful assessment of the benefits and costs of nonstructural options for improving waterway traffic should also be conducted.

Finally possible enhancements of environmental resources on the UMR–IWW may not have been adequately considered. The Corps should thus reconsider the issues regarding problem/opportunity definition on the UMR–IWW.

The role of Congress is very important in implementing this report's recommendations: full exploration of nonstructural approaches and proper inclusion of environmental assessments in the UMR–IWW feasibility study will require Congress to agree to this approach and studies, and then provide backing and adequate resources for the Corps to implement them.

Environmental Context of UMR–IWW Planning

The Corps' failed to consider explicit environmental improvements during the scoping processes in the UMR–IWW feasibility study. The Corps' draft study has been conducted within the context of several programs for improving environmental research and quality in the Upper Mississippi River basin, including the federal Upper Midwest Environmental Sciences Center (formerly the Environmental Management Technical Center) in La Crosse, Wisconsin, the interstate Navigation Environmental Coordination Committee (NECC), and the interstate Upper Mississippi River Conservation Committee (UMRCC). Both interstate groups consist of representatives from state environmental protection and conservation agencies in the five upper basin states (Illinois, Iowa, Minnesota, Missouri, and Wisconsin). The Corps has been and continues to be instrumental in these programs: to date, the Corps has allocated more than $181 million to the Upper Mississippi River System Environmental Management Program.

The Corps is not ignoring environmental considerations in the Upper Mississippi basin. However, the outputs of these interstate and federal initiatives have not been integrated adequately into a "systems" view within the UMR–IWW study. A true systems view of the UMR–IWW would acknowledge that a thorough understanding of the current state of the Upper Mississippi River and floodplain ecosystem must be related to structural modifications in the Upper Mississippi River basin, to operations of the UMR–IWW navigation system, and to ecosystem responses to a range of natural and anthropogenic influences within the Upper Mississippi River basin. The feasibility study would also benefit from a clarification of the roles of the environmental studies in the decision-making process regarding lock extensions.

ECONOMICS

The Corps' economic evaluation of lock extensions consists of two basic activities. First, the number of barges expected to use the lock is forecast, along with the congestion or delays that this traffic is expected to experience at the lock. Second, projections of traffic volumes and delays are then translated into estimates of the national economic development (NED) benefits of the lock improvement. The committee generally supports the NED theoretical framework that the Corps uses to estimate the benefits of lock improvements, but it has serious reservations about the methods used to forecast traffic volumes at the lock, and the empirical model used to estimate benefits.

Traffic Demand Forecasts

Waterway traffic demand forecasts are central to economic evaluation of proposed navigation improvements on an inland waterway system. The higher the volume of traffic that might use the waterway, and the less sensitive that traffic is to barge tariffs or costs, the greater the potential NED benefits from increasing waterway capacity (all other factors being held constant). If traffic volumes are growing and are relatively insensitive to barge tariffs, the waterway will become increasingly congested over time, and shippers, and the economy more generally, will benefit from capacity improvements. Conversely, the lower the traffic growth and the more sensitive shippers are to barge shipping rates, the lower the overall NED benefits.

As part of the UMR–IWW study, the Corps endorsed the use of spatial equilibrium models to estimate future levels of waterway traffic. Spatial equilibrium models recognize that freight flows represent a response to spatial variations in commodity prices and in the costs of transportation between points where prices differ. For example, Illinois corn is exported to Japan when the prevailing price of corn in Japan is high enough to cover the cost of buying the corn and then transporting it to Japan. The exports affect the supply and demand for corn in Illinois and Japan, driving corn prices up in Illinois and down in Japan.

In a key report of the UMR–IWW Economics Work Group (USACE, 1998), the Corps explained the advantages of a spatial equilibrium approach:

> The consensus of this literature is that the economic impacts of transportation systems are best analyzed as components of larger spatial price economic models. Analyzing transportation systems or their individual components myopically can lead to erroneous conclusions regarding economic impacts and values of the transportation system and its components. Spatial price economic models may be characterized as models where consumer demands and producer's supplies of goods and services are identified by their location in spatially geographic regions called markets.

The Corps' endorsement of the theoretical concept of spatial equilibrium is commendable, because accepted theoretical concepts will form a more credible basis for benefit estimation than the approaches formerly used by the Corps. Unfortunately, possibly because of limited time and data, the specific models the Corps developed for waterway traffic forecasts in the UMR–IWW study are inadequate. The models failed to forecast the correct direction of grain exports between 1995-2000. Moreover, the models contain no structural variables (underlying forces that determine directions and levels of grain exports; e.g., weather, income, government policies, changes in technology, and others) that engender confidence in the models' projections beyond 2000. Indeed, the shortcomings are so serious that the current results from the export forecasting model and the empirical ESSENCE model (used to model waterway traffic, levels of congestion, and changes in shipping rates) should not be used in the feasibility study.

Steps in Traffic Forecasting

To understand the Corps' waterway traffic forecast models, it is helpful to separate the various components typically involved in traffic forecasting. For simplicity, we use the example of corn shipments, because corn, soybeans, and other agricultural products account for an important part of the traffic on the UMR–IWW. The same framework applies, with some minor modifications, to other types of commodities typically shipped on a waterway.

The barge traffic forecast for any given year can be viewed as involving the following five steps, as shown in Figure 4.1 (Beesley and Kemp, 1987; Meyer and Straszheim, 1971; Ortuzar and Willumsen, 1994):

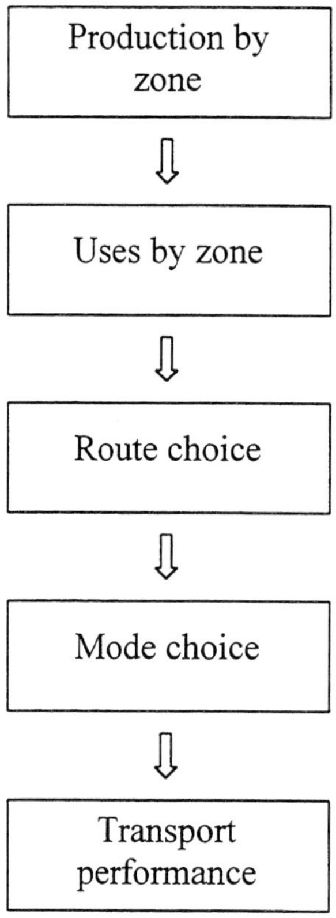

FIGURE 4.1 Typical steps in demand forecasting.

1. Forecast corn production in the area served by the waterway.

2. Forecast the uses and destinations of the corn. How much of the corn will be used as domestic animal feed, for example, and where are those pig farms located? How much will be exported and to which countries?

3. Forecast the routes by which the corn will be transported to its various uses. Will corn bound for export to Asia be shipped via New Orleans, for example, or via a West Coast port?

4. Forecast the mode, or combination of modes, that will be used to ship the corn on each route.

5. Forecast how the performance of each mode and route is affected by the traffic volumes carried. If large volumes of corn are moving through a waterway, rail line, or congestion may increase, pushing up costs and transportation tariffs.

These five steps are described here sequentially, when in fact they are nearly simultaneous. For example, an Illinois farmer would not consider planting an acre of corn without some idea of the price likely to be received. In turn, that price depends on how that corn is likely to be used and what the costs of shipping the corn to market are likely to be. Moreover, the steps must all be consistent with one another. For example, prices for corn in Illinois should reflect and be consistent with the costs of transporting corn to each market where it is sold. If the costs of shipping corn from Illinois were to rise significantly because of congestion on the UMR–IWW for example, Illinois corn might cease to be competitive on world markets. Illinois corn prices would have to fall to regain competitiveness in the world market, which would lower farm incomes and eventually lead to decreasing production.

Transportation forecasts usually have a high degree of spatial detail. The area served by the transportation facility is usually divided into zones, and estimates of production and uses (in steps 1 and 2) are provided for each zone. The analyses of the choice of routes and modes (in steps 3 and 4) and transport congestion and costs (step 5) are then done separately for each pair of zones that has traffic flowing between them. The more detailed the zone system, the more burdensome it is to calibrate and use the forecasting model. But because production, uses, transportation tariffs, and congestion can vary significantly over space, spatial detail is usually required to assure a reasonably accurate forecast.

Demand Models in the UMR–IWW Feasibility Study

The UMR–IWW feasibility study essentially divides the five steps described above into three groups (USACE, 1998):

- steps 1 and 2 (forecasts of total production of the commodity and its uses and destinations),
- step 3 (route choice), and
- steps 4 and 5 (mode choice and transportation system performance).

Commodity Production and Uses

Corn and soybeans bound for export are the major commodities carried on the UMR–IWW; the Corps developed two methods to estimate these flows. The first forecasts were based on projections of the amount of acreage along the waterway that would be planted in these crops and of the yield per acre of those crops (based on past years and projections). Estimates of domestic consumption of these crops were derived from historical trends in domestic consumption for various uses. The difference between production and domestic consumption was assumed to be exported. When this approach was shown to overestimate recent exports, the Corps' consultants (Jack Faucett Associates (JFA) of Bethesda, MD) instead adopted projections of agricultural exports for the region prepared by the U.S. Department of Agriculture (USDA) (JFA, 2000).

Route Choice

Spatial equilibrium models examine the routes between the origins and destinations of the commodities to be shipped. However, the major routing issue in the UMR–IWW draft study is whether the corn and soybeans exported from the study area will be exported through the ports on the Gulf of Mexico, U.S. West Coast or the U.S. East Coast. The Corps assumes that each port will retain its historic share of exports from each agricultural producing area. In practice, this means that almost all corn and soybeans exported from the UMR–IWW study area will continue to pass through the port of New Orleans.

A spatial equilibrium model answers these questions without the need to make an assumption about the share of each port. Having decided to use a spatial approach, it is unclear why the Corps decided to abandon the spatial model in favor of historical trends.

Mode Choice and Transportation Congestion

Forecasts of commodity flows by route are inputs into ESSENCE, a simulation model that forecasts how much of the traffic will be carried by barges, the levels of waterway congestion, and the resulting increase in barge costs. The ESSENCE model describes the performance of the waterway system in great detail, but it makes rather simple assumptions about alternative modes of shipping. The ESSENCE model has two main components[1]:

- a lock performance submodel that calculates delays and barge costs as a function of traffic levels at the locks, and
- a waterway market share submodel that calculates how much of the traffic moving in the corridor served by the waterway will be carried by barge.

ESSENCE is implemented using an Excel spreadsheet, and these two submodels are reconciled using Excel's "solver" function. The market share submodel assumes that certain towboat movements that do not pass through any locks are not sensitive to lock congestion. In addition, shipments moving by rail—where rates are below barge rates—are not sensitive to waterway congestion because they move by rail. For all other movements, the tonnage using the waterway in the forecast year, Q, is assumed to be a function of the relative tariffs for barge and the next-best-alternative mode:

$$Q = [(R - y) / (R - W_0)]^N * Q_0 \qquad (4.1)$$

In equation 4.1, R is the rate for the alternative mode, and the maximum that the barge company can charge, y is the estimated barge tariff, W_0 is the observed barge tariff in the base year (1994), Q_0 is the estimated tonnage that would be carried by barge in the forecast year if barge rates remained at W_0, and N is an empirically determined constant.

[1] There is a third component that calculates the net economic benefit from the waterway.

The sensitivity of waterway tonnage to barge tariffs depends strongly on the values of R and N (see Figure 4.2). R sets an upper bound on the tariffs that the barges can charge while N determines how sensitive levels of barge tonnage are to changes from the base year barge tariff, W_0. If N is greater than 1, barge traffic is very sensitive to changes in tariffs around W_0, but if N is less than 1, barge traffic is relatively insensitive.

Strengths of the Corps' Models

The demand models the Corps developed for the UMR–IWW feasibility study represent important advances over past Corps practice. In particular, using a systems approach in which improvements to all locks in the system are considered simultaneously, with consistent input data and with explicit consideration of project timing and other interdependencies, is commendable. Investment decision studies of comparable geographic and temporal scope are (unfortunately) rare for other freight transportation modes. Also, the recognition that a shipper's willingness to pay for navigation services is more complex than simple next-least-expensive mode calculations, and that it might even involve alternative markets or other types of business decisions, is an advance over previous methods. These theoretical developments are most welcome, and efforts to transform these concepts into useful decision support models should continue.

Weaknesses of the Corps' Models

The Corps' models should be improved in four areas: (1) the forecasting of domestic and foreign commodity production and use, (2) determining the sensitivity of waterway tonnage to barge tariffs, (3) calculating NED benefits, and (4) modeling lock congestion. All four areas affect the results of the evaluations of improvements on the Upper Mississippi River–Illinois Waterway, but the first two are particularly important.

Forecasts of Domestic and Foreign Commodity Production and Use

The Corps' initial forecasts (JFA, 1997), and the USDA projections (on which the Corps now relies) of U.S. crop production, consumption, and exports are based largely on simple extrapolations of past data. The initial projections of grain transported and exported were made in 1994. The 1995–2000 forecasts were incorrect; in fact, these forecasts called for increases in grain exports, but total exports of corn and wheat trended downward for this five-year period, with an expectation for continued declines in the near term. This led the committee to have no confidence in the forecasts beyond 2000. This model assumes that importing countries will buy all residual U.S. grain supplies remaining after domestic consumption. These exports were then allocated to importing countries based on historical shares.

The most recent projections used by the Corps are USDA projections (USDA, 2000) that are also based on extrapolating past production, consumption, and exports into the future. Nei-

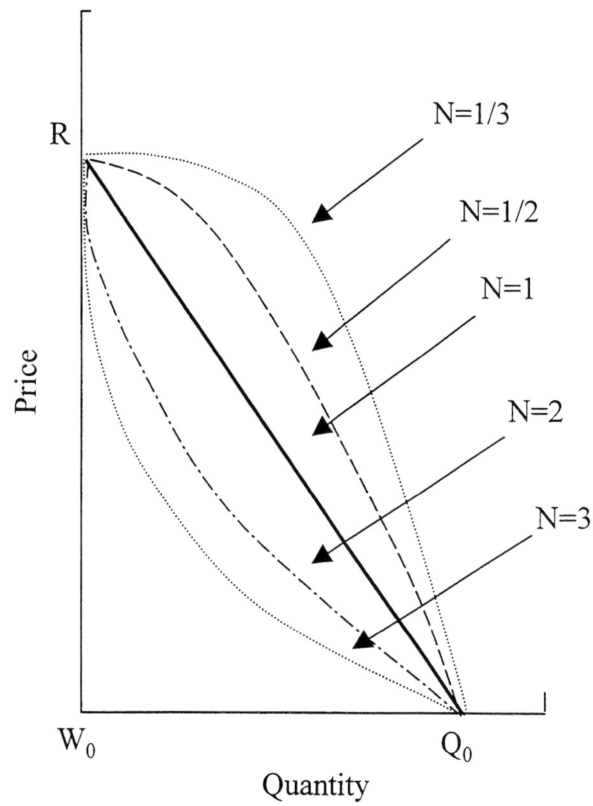

FIGURE 4.2 The Barge Shipment Demand Curve in the ESSENCE Model. SOURCE: USACE, 2000a.

ther the initial JFA projections nor the large-scale agricultural sector models—including the USDA model—adequately recognize world supply and demand factors. As a result, the export forecasts and projections do not fully recognize recent huge increases in world grain production. These forecasts underestimate the impacts of new technologies (e.g., biotechnology, irrigation, mechanical and chemical inputs) adopted around the world, the expansion of agricultural lands in South America and in the rest of the world, and the increasing resistance of consumers to genetically modified grains. These forecasts implicitly assume that all grain exports are in the bulk form (i.e., not packaged). Increasing exports of high-value grains and grain products are typically shipped in containers by rail to East Coast or to West Coast export ports and are not available for barge movements. Major exceptions include high-oil corn and white corn.

Models that base projections largely upon past trends, without adequately accounting for changes in supply and demand for grain in the rest of the world, do not provide reliable forecasts. The Corps' model failed to recognize shifts in importing country demands. For example, recent increases in meat exports from the United States to Japan, Taiwan, and South Korea are direct substitutes for U.S. corn exported to these three nations that are the largest importers of U.S. corn. This substitution is occurring as these three countries decrease the number of animals fed (because of outbreaks of contagious foot-and-mouth disease) and as feed efficiency increases in livestock production (a result of confined animal feeding operations and improved feed formulation).

Jack Faucett Associates was asked to determine whether the difference between the actual and initial forecasted exports was the result of short-term phenomena or long-term trends. JFA identified the short-term phenomena that affected the export volumes from 1995–2000. The Mississippi River floods of 1993, the loss of acreage planted because of the flooding, the closing of the river to navigation during the 1993 floods, and "hot spots" in grain shipments were all identified as short-term phenomena not likely to reoccur. The Asian financial crisis of 1997, an increased world supply of grain, and restrictions on grain exports to Europe were also offered as rationale for decreased or steady volumes of barge traffic.

But the fact that the fastest-growing market for U.S. corn is the domestic market, including uses such as wet and dry corn milling and increased usage of corn as animal feed (USDA, 2000), was not acknowledged. Furthermore, rapid growth in grain production in the rest of the world (because of increased yields per acre and increased acreage) is clearly reducing U.S. export growth potential. The evaluation of long-run versus short-run impacts seemed to ignore such trends.

JFA abandoned its initial method and projections and substituted extrapolations of USDA corn and soybean export projections. The basic premise of this attempt is inappropriate. In fact, in its initial report, JFA downgraded the USDA Baseline Projections to 2005 by stating, "The USDA further states in this publication that the projections are not intended to be a Departmental forecast of what the future will be, but instead a description of what would be expected to happen with an extension of 1990 agricultural law as amended. These statements are reflective of existing long-term materials to which comparisons might be made and also indicative of the problems in doing so" (JFA, 1997). Yet, JFA relied almost solely on this projection for its second set of forecasts. For years, the USDA has stated that their projections are not forecasts. Instead, they are conditional, long-run scenarios about what would be expected to happen under the 1996 Farm Act and specific assumptions about external conditions. This procedure ignores a downward trend in U.S. grain exports since 1980.

JFA's second set of forecasts is tempered by the supply and demand conditions offered by the USDA, even though these conditions are assumed to exist for only the next 10 years. However, the second set of projections for 50 years into the future was obtained by simply extrapolating the 10-year USDA projections an additional 40 years. An additional problem is that the forecast is affected by the choice of which data to use for the regression analysis. Moving the baseline year toward the present results in a forecast of low traffic growth. When an arbitrary decision as to which year to use as a baseline has a substantial effect on forecasts, one clearly cannot have confidence in the forecast.

There is no economic, statistical, theoretical, or empirical basis for extrapolating the USDA projections. A spatial equilibrium model, including detailed supply and demand functions for individual importing and exporting countries, is an integral part of this modeling analysis. This would likely require major modifications of existing large-scale agricultural sector models, supplemented by inputs from advisory groups and verified by sophisticated trend analyses.

Sensitivity of Waterway Tonnage to Barge Tariffs

Another serious problem is that there is little empirical basis for the Corps' estimate of how much of the grain will moved by barge, and how this will be affected by waterway congestion and the consequently higher shipping costs. The Corps' calculation is based on the ESSENCE model, which assumes an unusual functional form to forecast demand for barge transportation, as shown in Equation 4.1. The form is not necessarily wrong, but because it is unusual, it is harder to compare it to more conventional formulations and to understand the implications of different values of its parameters.

Corps economists originally argued that N in equation (4.1) was likely to be equal to 2, based on elegant but simple reasoning about the spatial distribution of commodity prices near a waterway. The economists hypothesized that agricultural products are loaded at a point on the waterway and that the point is surrounded by farmland producing a constant yield per acre. Each farmer has a choice between shipping the crop to export from the waterway port and selling it for local uses. If the costs of transporting the product to the port are proportional to the distance between the farm and the port and all farmers have the same alternatives, then the area shipping to the port will be a circle centered on the port. The size of the circle will expand or contract as barge rates at the shipping point fall or rise. Because the amount shipped on the waterway is proportional to the area of the circle, this implies that N is 2.

Although the reasoning is elegant, some of the assumptions used are not realistic. If loading points are close, for example, then the drawing area (the trade area in which transportation costs to a given loading point are lower than those to any other loading point) for each river port cannot be a circle, because the drawing areas will overlap—which implies a lower value of N. Similarly, the costs of transporting the grain to the loading point are not proportional to distance because the loading–unloading charge is fixed. Finally, the supply of grain is not uniform across all farms, and the farmers face somewhat different local alternatives. Theoretical issues aside, the important issue is that the Corps made no effort to assure that its assumptions about N were consistent with historical data on shipper behavior. Studies based on actual shipper behavior suggest that, contrary to the ESSENCE model, price responsiveness of freight demand varies greatly by commodity and by location (Small and Winston, 1999).

The ESSENCE model implies that farmers who choose to ship by barge lie within a circle centered on the loading port. As relative barge rates decline, the radius of the circle expands, since farmers further from the port can afford the shipping costs to the port, and vice versa. ESSENCE also implies that farmers close to the port (well inside the radius of the circle) are insensitive to price, since it is cheaper for them to ship by barge. Similarly, farms far away from a port (well outside the radius) are insensitive to price since they will almost never ship by barge. However, farmers that find that the cost of shipping by barge is almost identical to shipping by a different mode (or selling the corn for a purpose other than export) are extremely sensitive to small changes in price, such as that from one month to another.

Alternative Demand Curve Specifications, Willingness to Pay Concepts, and Calculation of NED Benefits

The ESSENCE model assumes that every shipment is sensitive to price, with an even distribution of willingness to pay (WTP) from the maximum willingness to pay (usually the alternative mode shipping cost) down to the market price of barge service. This contrasts with prior approaches, where each shipment that uses the barge mode is assumed to have the maximum willingness to pay. This maximum is usually assumed to be the alternative mode cost.

The Corps is to be commended for adopting a model where barge demand is modeled with users having a distribution of willingness to pay. However, this distribution is assumed rather than measured. In fact, the Corps has stated that no study resources were directed toward identifying the distribution of willingness to pay for commodity movements. This is a crucial omission. Obtaining data on the actual willingness to pay distribution is of high priority, because so much of the estimated benefit comes from the willingness to pay calculation. This benefit estimate should be based on data, not just unsupported theory.

Although the Corps endorsed the theoretical concept of a spatial equilibrium model with multiple origins shipping to multiple markets, linked together by commodity prices and transportation rates, these fundamental concepts are missing from the ESSENCE model. The only market in the ESSENCE model is New Orleans, Louisiana. The model contains no specific production areas, no alternative markets, and no market prices. Because there are no specified production areas, the model cannot link the production areas with alternative markets, including New Orleans. Thus, the ESSENCE model is not a spatial equilibrium model.

To conduct a meaningful analysis, it is important to include sub-regions in the study area that are connected by transportation costs/rates to various markets. One would expect each region to be linked by rates and prices to all alternative markets. When river traffic, congestion, and barge rates increase, it may be best to ship by alternative modes or to alternative markets. If a model does not allow for this to occur at numerous sub-regions throughout the study region, the resulting analysis will not be meaningful.

A major reason for the failure to include alternative modes, markets and routes, was the lack of appropriate data. Some of the problems with the data used in the grain export analysis in the UMR–IWW draft feasibility study are described below.

Origin Data. The ESSENCE model includes no information on specific origins (e.g., points from which grain is shipped). Rather, the quantities of corn, wheat, and soybeans are, in effect, piled at one location at each navigation pool. In the real world, the farmer makes the first decision that determines grain flows. Because there are no specific origins in the ESSENCE model, the impacts of the initial logistic decisions on lock extensions cannot be estimated.

Alternative Ports. Although the Corps allocates the historical shares of exports to other export ports, there are no alternative markets that include export ports in the ESSENCE model. Because the New Orleans market is the only port of export in the model, the model does not allow shippers to react to congestion by shipping to other markets. Because the cheapest way of relieving congestion may be to ship to other markets, ESSENCE probably tends to overstate the estimated costs of congestion.

Market Prices. Failure to include alternative markets in the model means that one of the two principal determinates of grain flows—prices—was not permitted to function in the model to determine grain flows.

Transportation Rates. The transportation rates used in the ESSENCE model are based on a confusing combination of rates and estimated costs (TVA, 1996). The Corps' Rock Island district office contracted with the Tennessee Valley Authority's (TVA) Water Resources Project and Planning Division to provide rate analysis for the UMR–IWW draft study. Estimated barge rates for grain and feed ingredients were correctly based on "percents of tariff" from the Waterway Freight Bureau Tariff 7 (this tariff was canceled, but the percentages of rates from this tariff have been the standard basis trading benchmarks since July 1979). However, the TVA "purged the data of anomalous or transitory influences. As part of all shipper surveys and interviews, respondents were directed to ignore temporary market disruptions and provide information reflective of 'normal' operating conditions" (TVA, 1996).

These comments notwithstanding, the only normal thing about barge rates is their variability. The Tennessee Valley Authority, however, removed all variability in the barge rate data. In reality, barge rates generally vary from 80 percent to 400 percent of the tariff (Merchants Exchange of St. Louis, 1979). Except for areas close to the Mississippi River, barge rates above 180 percent are typically higher than rail rates (ibid.). These high barge rates often create huge temporary demands for rail cars. A classic example of high barge rates and rail car shortages occurred in the last half of 1995 and the first quarter of 1996. Barge shipping rates were higher than rail shipping rates during this entire period. Because barge rates are usually lower than rail rates, removing the barge rate variability essentially assumes that rail rates will always exceed barge rates. Because actual barge rates are often above rail rates, an assumption that this is not the case will result in overestimating the NED (since the assumption will divert actual traffic from rail to barge).

Railroad rates were obtained from a nonrandom sample of shippers and receivers, apparently mostly barge shippers and receivers. However, because much of the rail rate data were incomplete, rail rates were also estimated using the Reebie Rail Costing Model (Reebie Associates, 1999). These estimates were "inflated to reflect rail carrier market power in order to produce a final estimate of the most likely tariff" (Reebie Associates, 1999). However, the TVA report did not specify what it meant by market power or what procedures were used to adjust the rail rates used in the ESSENCE model. This problem might have been avoided by extracting railroad rates from published railroad tariffs available on the Internet.

The TVA report further states that the lower of the tariff rate or the estimated rail rate was actually used in the analysis (the tariff rate is a publicly-posted price available to all takers who are able to meet conditions specified in the tariff, such as size and time of shipment, on a take it-or leave it basis). However, it is not clear how the TVA could compare estimated rates with "incomplete information" rate data to select the appropriate rail rate for inclusion in the model. Moreover, the TVA analysis excluded new railroad technologies in the railroad rate analysis. For example, the TVA excluded contract rates (the dominant rail rates used in much of the U.S. Midwest), guaranteed car supply agreements, shuttle and cycle train rates, new and larger covered hopper rail cars, and new highly fuel-efficient alternating current railroad locomotives. Failure to include these technical and pricing innovations biases rail rates upward and overestimates benefits.

Matching Rates and Quantities. The greatest problem with the data on quantity (of grain shipped by barge) and rates (rates charged for those shipments) is that the quantity data are from 1991 and the rates charged are from September 30, 1994. Thus, the quantity and rate data used in the ESSENCE model do not match. The 1991 quantity data are the result of a different set of rates than those used in the model, and the 1994 rate data generated a different set of quantities than the 1991 quantities used in the model. There is thus no meaningful relationship between the quantity and rate data. In this instance, the 1991 and 1994 data were significantly different, but this is beside the point. The point is that a demand curve cannot be properly constructed with price and quantity data from different years.

Lack of Intra-annual Rate Variability. Barge and rail transport rates, and quantities shipped, often vary enormously within and between years. Yet the rate data used in the ESSENCE model are for one day—September 30, 1994. Given the huge variability in rates and quantities within years, the data should have been collected for, and the model run for several time periods within each year. Failure to allow for rate variability eliminates the possibility of barge rates exceeding rail rates in the model and biases the calculation of NED benefits upward.

Absence of Spatial Data. A major reason for the absence of spatial equilibrium characteristics (e.g., no specific origins of grain in the model) in the ESSENCE model was that the Corps did not identify and estimate supplies of grain available in local areas. The Corps also did not:

- collect rate data from local supply regions to alternative markets,
- collect price data at alternative markets to enable local supply regions to select the most profitable markets at different time periods within the year, and
- specify maximum quantity constraints at nonbarge alternative markets.

Nonagricultural Commodities. As is true for agricultural shipments, the ESSENCE demand equation was used to forecast waterway transport demand for nonagricultural commodities as a function of the alternative mode rate and the water transportation rate. The two basic inputs to this process are the other mode cost (i.e., rail) and the demand curve N value (see equation 4.1). Other-mode costs were estimated by TVA, based on in-depth analysis of a sample of existing UMR–IWW shipments. For each shipment, TVA relied on industry data and published rates, supplemented by a total cost assessment that included load/unload costs, linehaul rates, and handling costs. In all cases, rail, truck, or combination moves were analyzed as the other mode, with allowances for modal access conditions at each end of the existing water move. In some cases, combination land-water moves, or moves to alternative destinations, were analyzed as the next-lowest-cost alternative. The sample results were used to develop linear regression models for extrapolation of the sample results to the population of all waterway movements. Unfortunately, the data used in the mode and route analysis for nonagricultural products suffer from many of the problems encountered in the grain analysis, including:

• Estimated barge costs, rather than actual barge rates, were used. Fewer than 35 barge companies operate covered hopper barges (CHB) on the Mississippi River system, and five of these firms operate almost 70 percent of the CHBs. One of these companies operates about a third of the CHBs. These market shares indicate a moderate level of concentration for barge transport. Rail transport operations are more concentrated. While there are approximately 350 railroad companies in the U.S., about 95 percent of the nation's grain shipped a significant distance is carried by the top five large, Class I railroad companies. This high concentration in the railroad industry and the moderate concentration in the barge industry means that neither industry is close to being purely competitive. Long-run marginal costs are therefore not a reasonable proxy for rates.

• Barge companies may be comparatively few (and continue to decline to number), but a particular barge-loading elevator typically has access to more than one barge company on the Upper Mississippi River. While inland rural grain elevators typically have access to only one railroad (if at all), they have access to many trucks (trucks are today have the largest share of grain shipments of all three modes). This certainly provides major competition to those railroads that serve inland country grain elevators. Rail rates were estimated from the Reebie Rail Costing Model plus a markup based on shipper information and/or the Interstate Commerce Commission (ICC) Waybill Sample data. It is not clear how shipper information could be used to determine railroad markups. Moreover, the ICC and the Surface Transportation Board (the successor to the ICC) specifically state that waybill data should not be used to make rate comparisons because of disguised rail rates in the sample. Grain sales and shipment decisions are made on the basis of commodity prices and transportation rates, not long-run railroad, barge, and truck marginal costs. Although long-run marginal costs may approximate truck rates because of the large number of trucking companies, they certainly do not approximate rail or barge rates.

• Rate and quantity data that do not match temporally. The quantity data are based on a sample of 1991 shipments whereas the rail and barge rate estimates were for September 30, 1994. This distorts the demand curve for barge traffic. A meaningful economic analysis requires that the rate and shipment data be for identical time periods.

• Exclusion of technological innovations and the impacts of regulatory changes on rail operations and pricing biases rail rates upward.

• No alternative origins and/or destinations are considered.

• No transportation rates to or from the alternative origins or destinations are considered.

Another important issue is whether the analysis properly captures actual alternatives faced by waterway shippers. If the waterway were not available, would shippers shift to the next-lowest-cost mode, and if so, would they continue to ship the same quantity? If this truly is the shipper's second best choice, then the other-mode cost is the willingness to pay, and the NED benefit of the waterway is the applicable consumer's surplus (i.e., willingness to pay minus water equilibrium price). However, the shipper has other alternatives: ship the product to another market or sell it for another use, produce less, or even go out of business if the other-mode cost is too high. The highest net revenue from these other responses defines the actual willingness to pay for this shipper. Again, this is an empirical question. All real alternatives to shipping by water have a net revenue that can be estimated.

There are some data, based upon historical experiences with lock chamber closures for rehabilitation projects, that shippers will continue to ship by water in the face of higher prices, and that shippers will shift to the next-higher-cost mode in order to keep shipments arriving at their production or export facilities. However, these are short-run responses, and there are no known long-run responses, as data on shipper impacts were not collected in those few instances where a waterway was entirely or partially closed to commercial navigation.

The data requirements for the draft navigation feasibility study are large and could be very costly. A crucial aspect of planning the feasibility study is for the Corps to allocate its resources carefully among the study components and to assess whether to request additional funding because important data cannot be acquired with current funding. In many instances within the navigation study, the Corps made good decisions about allocating study resources. In some instances, however, data were gathered that were of little use in evaluating the benefits and costs of lock extensions. For example, the rate and quantity data for different years were of no use in the study, as no inference can be drawn from these data. Much of the data collected for the environmental studies were also of very limited value: while some of the individual environmental studies may have been scientifically sound, it is not clear how the results from those studies affect the decision regarding lock extensions. This committee was not charged to design the best way to conduct a $50 million study, nor were resources made available to do so. We can only point out that the Corps did not always spend its money wisely and specify the additional data that are needed.

Modeling Lock Congestion

The Corps used a relatively simple queuing model to represent the increase in delays at UMR–IWW locks in response to traffic increases. However, most of the Corps' district and division offices and research centers have determined that simulation modeling (rather than queuing theory) is the preferred approach to analyzing lock performance, and that it can account for the mix of vessel types and operating phenomena that occur. Several such models are available (e.g., Dai, 1993; USACE, 1993). With a series of locks 10–30 miles apart, and with most traffic attempting to pass through these locks without interruption, random arrivals assumed in the Corps' queuing model do not match the actual traffic pattern. Finally, simulation models are much better suited for determining benefits of operational improvements (such as lockage scheduling or industry self-help), as suggested elsewhere in this report. It is unlikely that a queuing model will be able to provide good delay-reduction estimates for these types of measures.

Treatment of Uncertainty

The inherent uncertainties in forecasting barge shipments up to half a century into the future are well known (Box 4.1 describes problems caused by placing too much faith in waterway traffic projections). Without treating uncertainty explicitly, the Corps' analysis might conclude that the benefits of extending locks on the UMR–IWW are 10 percent greater than costs,

BOX 4.1
UNCERTAINTIES AND COMPLICATIONS OF WATERWAY TRAFFIC FORECASTS

Waterway Traffic Projections on the Mississippi River: The Lock 26 Replacement Project

The Lock 26 replacement project, which constructed Melvin Price Lock and Dam, is a mixed story in terms of the ability to forecast waterway traffic levels. Table 4.1 compares the actual traffic passing through Melvin Price's two lock chambers in 1999 to two projections, the first made in 1975 and the second made in 1982.

TABLE 4.1 Projected Traffic Flows Through an Expanded Lock 26 vs. Actual 1999 Traffic Flows

	1999 Projected Flows		
Commodity	1975 Projection	1982 Projection	1999 Actual Flow (million tons)
Grains incl. Oilseed	44.45	79.22	44.05
Coal	16.53	13.15	7.09
Petroleum and Products	34.99	8.50	3.32
Industrial chemicals	9.84	13.91	2.16
Other	19.97	8.58	20.96
Total	123.79	123.16	77.58

Source: 1975 Projected traffic flows compiled from Corps of Engineers Lock and Dam 26. Replacement Design Memorandum No. 11, Vol. 1, Table 3-52, as quoted in U.S. Department of Transportation, The Replacement of Alton Locks and Dam 26: An advisory Report of the Department of Transportation to the Senate Commerce Committee, 1975, p. 13. The 1999 figures are interpolated from 1985 and 2000 traffic projections. The 1982 projection from Technical Report A, Navigation and Transportation, Comprehensive Master Plan for the Management of the Upper Mississippi River System, prepared for the Upper Mississippi River Basin Commission by U.S. Department of Army, Corps of Engineers. Figures are an interpolation of 1990 projections from Table 1.8 and 2000 projections from Table 1.9. The 1999 actual tonnages are from Corps of Engineers, U.S. Waterway Data, Lockage Statistics.

Table 4.1 shows that both the 1975 and the 1982 projections greatly overestimated the amount of traffic that actually passed through Lock 26 in 1999. Traffic was forecast to be 63 percent higher in 1999 than it actually was.

The source of the overestimate of traffic is different between the two years. The 1975 projections of grain movements through Lock 26 were absolutely on target, with 44 million tons of grain projected to move through the expanded lock and 44 million tons actually moving through the main and auxiliary chambers.

The reason that the 1975 total traffic flow forecast missed the target so badly was that projections of other types of traffic were not as accurate. A particular disappointment is that only one-tenth as much petroleum and petroleum products actually moved through the system as expected. In fact, only half as much petroleum moved through Lock 26 in 1999 as moved through the old unexpanded lock in 1966. Industrial chemical movements also declined after 1966 rather than increasing, as was expected. Coal movements in 1999 were at the same level as 1969, rather than more than doubling, as forecast in the original traffic projections.

Projections for the Tennessee–Tombigbee Waterway Project

The Tennessee–Tombigbee waterway project provides an example of overly optimistic traffic projections. The figure below shows projected and actual traffic levels since the canal was opened in 1986. Waterway traffic has fallen far short of what was projected. In recent years, the difference between expected and actual movements amounted to 80 percent of the original forecast tonnage.

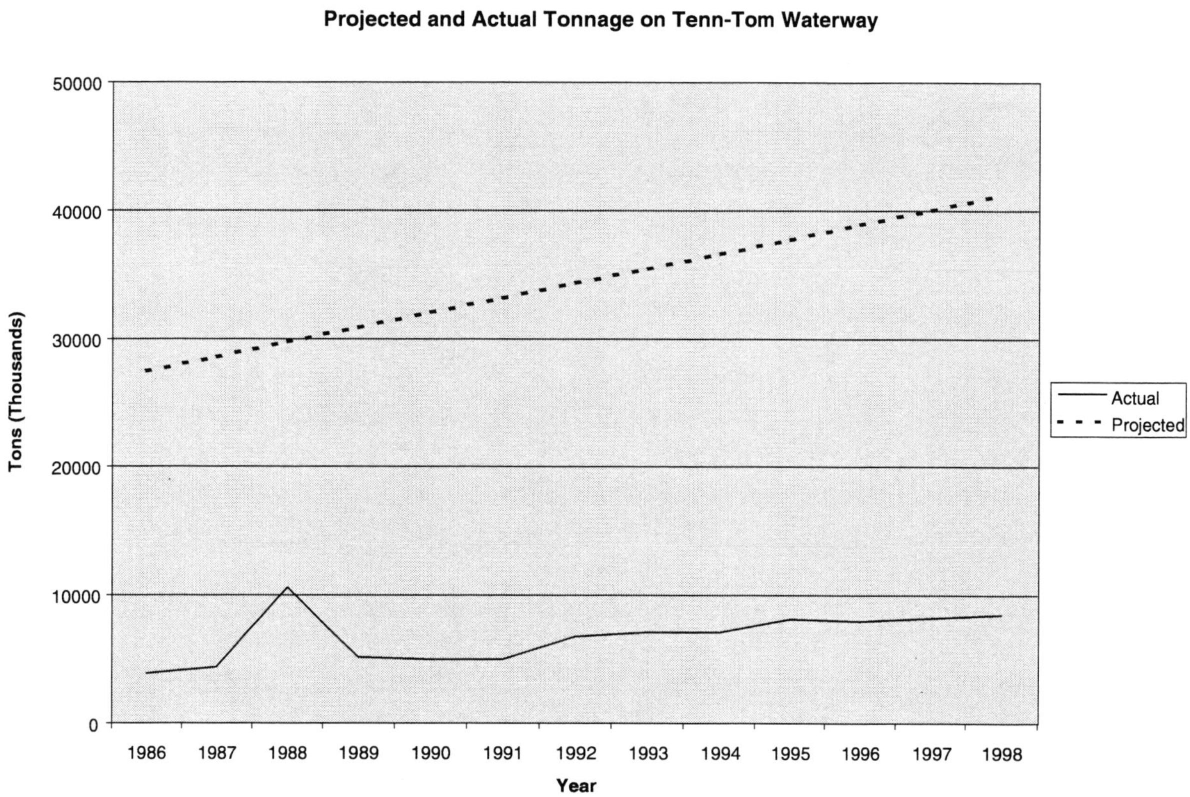

The primary reason for these highly inaccurate forecasts is that coal shipments did not develop as anticipated. The primary rationale for building the canal was to move coal from mines in Eastern Kentucky to the Gulf of Mexico. The benefit–cost study used to justify the building of the canal forecasted coal production in eastern Kentucky to increase several fold in response to the high energy prices of the 1970s. That new traffic was then expected to travel the shorter route to the Gulf of Mexico through the new canal rather than the traditional route via the Tennessee, Ohio, and Mississippi Rivers.

At the time of the original study, this forecast was criticized by some transport economists (Carroll and Rao, 1978). At that time, there was reason to doubt the forecast's assumption that high oil prices would lead to massive increases in coal production—especially the high-sulfur variety found in eastern Kentucky. In fact, U.S. coal production has increased modestly since the 1970s, but the increase has been in the production of low-sulfur coal in the western United States. In addition, transport economists noted that the Tennessee–Tombigbee navigation project, while presenting a shorter route to the Gulf, was not a less costly route to the Gulf. And in fact, the much larger tows allowed by the broad

Lower Mississippi have attracted the limited amount of Kentucky coal traffic to the more circuitous route. In the project's first year of operation, only 500,000 tons of coal were shipped on the canal, rather than the 17 million tons that had been forecast. In 1998, coal tons on the Tennessee–Tombigbee had increased to 2,500, but the shortfall between actual tons and projected tons (which were expected in the original traffic projection to grow rapidly) continues to increase. The main commodity shipped on the Tennessee–Tombigbee waterway today is wood chips, not coal.

Criticisms of the Tennessee–Tombigbee navigation project rest squarely on overly optimistic traffic projections. In order to avoid such problems in the future, waterway traffic projections for new projects should be examined far more carefully to eliminate the possibility of upwardly biased forecasts.

and that construction needs to begin in 2001 in order to help grain farmers avoid the high barge rates that will make them less competitive in the world market. An uncertainty analysis might lead to the following conclusion (this is purely hypothetical, as the committee did not conduct its own uncertainty analysis): there is a slightly greater chance (55 percent) that benefits will exceed costs, rather than the reverse (45 percent). Greater information about the nature of the uncertainty comes from the distribution: it is somewhat likely (10 percent probability) that lock extensions would have costs 70 percent greater than the benefits, or that benefits would be 20 percent greater than costs. The extended locks might have annualized benefits greater than annualized costs as early as 2015 or as late as never.

For example, the initial analysis done by the Corps relied on forecasts by Jack Faucett Associates for the period 1995–2045 (JFA, 1997). However, the first five years of this forecast—the period in which forecasts would be expected to be the most accurate—show falling, rather than rising, exports. In an analysis that does not treat uncertainty explicitly, this forecasting failure is shocking and casts doubt on the entire analysis. If the uncertainty analysis encompasses the possibility of declining use during a period, the analysis would be more credible.

Another example is the construction cost of Lock and Dam 26 (Melvin Price Lock and Dam) at Alton, Illinois. Originally budgeted to cost $350 million, the final cost of Lock and Dam 26 was $1.05 billion. The Corps has a standard contingency of 25 percent for cost overruns. Without an explicit uncertainty analysis concerning costs, Congress might conclude that the Corps cannot do competent engineering, or at least cannot estimate costs competently. The committee understands that other factors entered to increase the costs almost 3-fold. However, for large construction projects, particularly ones that involve environmental considerations, delays and project changes are likely and costs almost inevitably increase. An explicit cost analysis informs Congress and other parties about what might occur. If setting out the uncertainties explicitly erodes support for the project, perhaps the project should not have been built.

The fault lies not in getting the forecasts wrong. It is impossible to accurately forecast barge movements 50 years into the future or to forecast the costs of a large construction project. Rather, the fault lies in not treating the forecasts as highly uncertain. In Chapter 5, we set out the current state of the art in modeling the uncertainty associated with forecasting future events. Comparing the year 2000 actual barge movements with the forecast movements, or the forecast costs of building Lock and Dam 26 with actual costs, is humbling for the analyst. We point out the differences not to imply that Corps analysts are deficient, but rather to point out the inherent uncertainty in these forecasts. The differences point out the need for explicit treatment of uncertainty, with full recognition that it is impossible to predict the future with precision. Rather than setting out precise values for future barge movements and other relevant events, the Corps should have examined the uncertainties explicitly, commenting on the range of values and the conditions under which demand was high enough to lead benefits to exceed costs.

ENVIRONMENT

In 1991, following the initial reconnaissance studies conducted on the Illinois and Mississippi rivers, the Corps determined that the UMR–IWW navigation draft feasibility study would take a system-wide approach in addressing navigation problems common to both the Illinois and Upper Mississippi waterways. However, instead of examining adverse environmental conditions created by the existing navigation system and proposed improvements, the Corps chose to focus narrowly on only those effects associated with future increases in barge traffic. However, it bears repeating that the 1970 Flood Control Act does not require the Corps to seek improvements in environmental quality when it considers changes in existing project operations. Rather, the 1970 act authorizes the Corps to seek these improvements as they see fit.

In the early 1990s, there were requests to broaden the scope of the environmental studies, including a request from the Navigation Environmental Coordinating Committee to assess the long-term impacts of operating and maintaining the navigation project on the Upper Mississippi River. The Corps' justification for limiting its environmental analysis to incremental effects of additional waterway traffic was apparently fiscal in nature: "A broader multipurpose environmental effects study proposed by a number of agencies and organizations would require a 50/50 cost sharing by the states or other sponsors" (USACE, 2000b, p. 9).

In the Corps' draft navigation feasibility study, environmental components were developed to meet planning objectives consistent with the Corps' analysis of the navigation system, and consistent with the analysis of incremental effects of increased barge traffic on river ecology. The primary environmental objective was to assure that any recommended actions were consistent with protecting the nation's environment by avoiding, minimizing, or mitigating significant environmental, cultural, or social impacts. This objective was subject to a planning constraint that directed environmental investigations specifically to assess impacts of the construction and to assess incremental traffic effects of any potential navigation improvements (USACE, 1997b). Other environmental investigations and studies also were considered separate from the draft navigation feasibility study.

Environmental studies tasks in the navigation feasibility study included (1) developing data management/mathematical modeling strategies for extrapolation of impact data to the Upper Mississippi River system, (2) analyzing physical effects of field data previously collected by the Illinois State Water Survey, (3) conducting physical effects studies at the Corps' Waterways Experiment Station in Vicksburg, Mississippi, (4) evaluating navigation traffic effects on sedimentation of side channels and backwaters and on bank erosion, (5) developing a geographic information system (GIS) database covering representative river reaches and habitats, (6) evaluating navigation impacts on fish, wildlife, and biota, including plants and mussels, and (7) comparing the impacts of recreation craft with commercial navigation impacts (USACE, 2000b). Fiscal constraints required prioritization of study efforts and elimination of lower-priority studies (Figure 1.2 illustrates cost estimates for the various components of the feasibility study).

The feasibility study has been conducted on a river system that has been the subject of many environmental studies and assessments. There is a long history of studies for both the Upper Mississippi and Illinois rivers (Belrose et al., 1979; Galloway, 2000; Hart, 1895; Nelson et al., 1998; Scarpino, 1985; Sparks, 1977). For example, the Upper Mississippi River Basin Commission conducted studies and developed a master plan to help the U.S. Congress resolve conflicts between commercial navigation and other groups in the region (UMRBC, 1982). The Upper Mississippi River Basin Commission master plan recommended a habitat rehabilitation and enhancement program, a long-term resource monitoring program, and a computerized inventory and analysis system.

The 1986 Water Resources Development Act (WRDA) responded to these UMRBC recommendations by establishing the Environmental Management Program (EMP) for the Upper Mississippi River. As discussed in Chapter 1, two of the main programs within the EMP are habitat restoration and ecosystem monitoring (the latter is known as the Long Term Resource Monitoring Program, or LTRMP). The Corps is responsible for habitat restoration, while the U.S. Geological Survey (USGS) coordinates the LTRMP (in 1996 ecosystem monitoring and other biological services were transferred from the U.S. Fish and Wildlife Service to the U.S. Geological Survey). As part of the Environmental Monitoring Program, the Upper Midwest Environmental Sciences Center in La Crosse, Wisconsin, monitors water quality, sedimentation, fish, vegetation, and invertebrates, as well as land cover and land use (USACE, 1997a; USGS, 1999). In a 1997 Corps of Engineers evaluation, the EMP was described as "fundamental to successful comprehensive management of the system" (USACE, 1997a). The Corps furthermore stated, "The EMP has come to be the single most important and successful program authorized by the Federal government for the purposes of understanding the ecology of the [Upper Mississippi River System]" (USACE, 1997a).

Lock and Dam 26

The authorization, construction, and environmental effects of Lock and Dam 26 (also known as Melvin Price Lock and Dam) at Alton, Illinois, figured prominently in the establishment of the Environmental Management Program, as well as in the establishment of several environmental initiatives leading up to the navigation feasibility study.

In the 1970s, a proposal to replace and increase the navigation capacity of Lock and Dam 26 generated a great deal of controversy. Lock and Dam 26 had one 600-foot lock and a 360-foot auxilary chamber; the proposal was to replace these with a new dam and with two 1,200-foot locks. In 1978, Congress authorized construction of one 1,200-foot lock and requested the Upper Mississippi River Basin Commission to conduct a study and make recommendations regarding further navigation expansion and environmental implications. The UMRBC Master Plan recommended (among other things) that Congress authorize a second, but only 600-foot, lock.

The 1986 WRDA, however, contained authorization for a second 1,200-foot lock at Lock and Dam 26 (as mentioned, the 1986 WRDA also authorized the EMP). To assess the environmental effects of this second lock, the Corps created a Plan of Study (POS), which was designed with participation of several agencies (see Appendix A for a list of participating agencies). The POS called for a 10-year study of the second lock's environmental effects, with an estimated cost of $27 million (USACE, 1991).

In its 1997 Project Study Plan, the Corps stated that the UMR–IWW navigation study would take no further administrative actions on studies of the effects of the second lock at Melvin Price (USACE, 1997b): "Concerning the Second Lock Plan of Study, no further administrative action will be taken. The environmental studies and analyses under way as a part of the UMR–IWW System Navigation Study will provide a rational basis for decision making."

Results from the study could have greatly improved understanding of the environmental effects of lock construction. These results would have been especially valuable as input to the ongoing feasibility study. Unfortunately, the second lock study at Lock and Dam 26 today merely represents a foregone opportunity to enhance understanding of human impacts on the Mississippi River system.

Environmental Data, Modeling, and Related Analyses

The environmental analyses in the navigation feasibility study are supported by the Corps' draft environmental impact statement (DEIS). The feasibility study also draws heavily from the environmental analysis in the Corps' 1997 Project Study Plan, which used information as part of the federal National Environmental Policy Act (NEPA) process and also used information from ongoing studies conducted within the EMP's Long-Term Resource Monitoring Program.

The foundation for the Corps' draft environmental impact statement was a risk analysis. The first step in this risk analysis conceptual model outlining the nature and sources of stress to ecological resources, ecological resources potentially at risk, ecological impacts of concern regarding these resources, relevant data and information, and suggested models and methods of analysis for estimating risks. Risk assessments were developed for fish, aquatic plants, and freshwater mussels. A report summarizing ecological models and the approach to ecological risk assessments (Bartell et al., 2000) was also issued.

The draft EIS examines only the effects of incremental increases in barge traffic, reinforcing the Corps' choice to limit its studies (dating back to 1991) to the effects of increased barge traffic, and not including the long-term effects of the operation and maintenance of the entire navigation system. As a result of this limitation, the detailed plans for analysis included only the ecological risk posed by commercial vessels to river banks (through erosion), and to submerged aquatic vegetation, freshwater mussels, and fish. The conceptual model developed to support the risk analysis (Bartell et al., 2000) was therefore limited and did not address the full scale of risks to the UMR–IWW ecosystem. The risk assessment design focused on specific biological effects such as early life stage mortality of fish (e.g., through direct entrainment of fish larvae into the propeller jets of commercial vessels), degradation or loss of fish spawning habitat, physical breakage of submerged aquatic vegetation, impacts on the growth and reproduction of submerged aquatic vegetation, and impacts on the growth and reproduction of freshwater mussels. Potential ecological risks posed by incremental increases in commercial traffic were then estimated using models that quantified the magnitude, extent, and duration of the physical forces produced by commercial vessels and using models that quantified the ecological effects of such physical forces. Alternative traffic scenarios provided input data (e.g., vessels per day, and vessel and barge configuration, direction, speed, and draft) for the physical forces models used in the Corps' risk assessment. Results from the physical forces models were used as inputs to the ecological models used to estimate the corresponding impacts for each traffic scenario. The general findings were that (1) during any site-specific construction activities, some habitats would be lost, (2) species- or reach-specific impacts are not expected to negatively affect sport fish abundance or catch, (3) with the exception of paddlefish and sturgeon, no potential impacts to the commercial fishery are anticipated, and (4) that since operations and maintenance practices would remain the same, it is not envisioned that any permanent changes would occur to rivers in the system (USACE, 2000b).

In addition to the shortcomings mentioned earlier, the committee found two major flaws of the UMR–IWW feasibility study documentation: (1) a comprehensive evaluation of navigation-related effects in the UMR–IWW has not yet been conducted and (2) although it is conducting a system wide study, the Corps has failed to complete a systematic study aimed toward integrating engineering, economic, and environmental issues.

Deficiencies of the Navigation Feasibility Study

Some groups (e.g., the U.S. Fish and Wildlife Service) assert that the current Corps analyses are deficient because despite the studies and progress made under the EMP, a comprehensive evaluation of navigation-related effects is incomplete. For example, in an August 31, 2000 letter the U.S. Fish and Wildlife Service argued that comprehensive evaluation would examine both direct and indirect effects of passing towboats as well as the effects of operation and maintenance (Appendix B). The U.S. Fish and Wildlife Service further noted that a system wide study should be conducted to evaluate cumulative effects. In that letter, five major deficiencies in the Corps' impact analysis were listed: (1) the assessment of incremental traffic effects is inadequate, and site-specific assessments are incomplete, (2) an assessment of baseline traffic effects in the "no action" alternative is needed, (3) an assessment of impacts from operation and maintenance activities is needed, (4) an assessment of traffic impacts and mitigation for the Second Lock (Lock and Dam 26) was never completed, and (5) a comprehensive mitigation plan that addresses all effects of the 9-foot channel is lacking.

Other critiques of the feasibility study stem from comparisons with the 1997 Project Study Plan studies and with studies recommended by the interagency team in the 1991 Plan of Study for the second lock at Lock and Dam 26. The Second Lock Plan of Study (USACE, 1991) was the outcome of state and federal agency cooperation encouraged by the UMRBC in its master plan. The proposed studies included 15 work units, seven of which were given highest priority. The high-priority work units included analysis of (1) basic physical forces, (2) side channel/ backwater sedimentation, (3) physical and biological effects of traffic in representative backwaters, (4) effects of navigation traffic on adult fish, (5) impacts on early life stages of fish, (6) navigation effects of suspended sediments on aquatic plants, and (7) navigation impacts on freshwater mussels.

The 1991 Second Lock Plan of Study has not been implemented, but elements were included with the UMR–IWW Project Study Plan (USACE, 1997b), which had a narrower focus and called for study completion in six rather than ten years. One outcome of this narrower study focus is that the possibility of gathering biological data necessary to support comprehensive river management decisions was greatly reduced. Reflecting some of the difficulties in biological assessment, there was a shift in the project study plan's environmental studies from an assessment necessary to improve understanding of both short-term and long-term biological conditions and ecological interactions, to the collection of physical data, such as hydraulic and hydrodynamic effects of towboat passage. Although these physical data have been useful in impact modeling and risk assessments, model application to the entire UMR–IWW system is still limited by the inadequacy of the biological and ecological data needed for model calibration.

Even though the EMP has improved the ecosystem database, the ecological information needed for a full assessment of model sensitivities, and to evaluate risk predictions in systemwide impact analyses, is not available. For example, while knowledge of river hydrodynamics has been improved, there are only very limited data to adequately model—on a systemwide basis—relations between river hydrodynamics, towboat passage effects, and biological responses. The lack of this type of interdisciplinary, systemwide understanding is a weakness of the draft navigation feasibility study and reflects the fact that no systemwide study of the long-term effects of the navigation system and its operations—including the ecological effects of the creation of a system of navigation pools—has yet to be conducted. A systemwide study assessing the operation and maintenance impacts associated with the Upper Mississippi River and Illinois Waterway system project should be conducted.

Lack of Systematic and Integrated Analyses

Although the Corps' navigation system feasibility study is ambitious, it does not provide adequate information on which to base important environmental decisions. Furthermore, the documentation provided to date does not effectively integrate the results of wide-ranging project study plan studies. The Corps has addressed systemwide issues in the draft navigation feasibility study, but as mentioned, it has yet to complete a systematic study of the UMR–IWW that effectively relates continuing ecological responses from past navigation system improvements with incremental effects and construction impacts in a realistic context that recognizes the influences of land uses in the watershed.

A common thread in criticisms of the navigation feasibility study environmental analyses (e.g., from the NECC and USFWS) is that individual elements of the feasibility study remain disconnected from the environmental analysis. This critique focuses on the lack of integration across engineering, economic, and environmental findings that would provide a systematic assessment of the long-term environmental effects of the navigation system's operation and maintenance. For example, the study of the effect of physical forces on fish or plants appears not to address the consequences of (a) spills from passing towboats, (b) indirect effects of habitat alteration on food or space limits in target and nontarget fish and mussel species, or (c) cumulative effects of multiple factors associated with increased barge passage, increased risk of spills, and increased operations and maintenance of the lock system. The absence of integrative analysis in the draft feasibility study to date is a concern.

In summary, the UMR–IWW draft navigation feasibility study, although somewhat of a landmark in the history of contemporary Corps of Engineers water resources project planning studies, has not conducted the environmental analyses necessary to provide information on, as stated in the 1970 Flood Control Act, "improving the quality of the environment in the overall public interest." The focus is on environmental studies that consider only incremental traffic effects, and not on a more comprehensive analysis of system-level consequences, and studies directed to the improvement of environmental quality. Furthermore, the draft feasibility study sought to avoid, minimize, or mitigate significant environmental impacts, which addresses only specific actions in specific locations, and left system-wide assessments to other efforts under different authorities.

ENGINEERING

Approximately 25 percent of the cost of the feasibility study supported the efforts of the engineering work group. This work group evaluated the current navigation system and the anticipated without-project operations and maintenance, rehabilitation, and replacement needs. It also conducted engineering and cost-estimating efforts to support plan formulation activities and evaluation of alternatives. The committee's review of this work was limited to examining whether the work provided an adequate basis for the economic assessment of the proposed system improvement alternatives. That is, the committee did not closely review the engineering analyses performed, but did consider whether the scope and depth of the analyses were consistent with efforts normally required at the feasibility study stage, and whether the engineering judgments and the conclusions reached were supported by the data displayed in the draft report.

Lock and Dam Rehabilitation

For some alternatives, a significant portion of the benefits of lock chamber extensions comes from savings in anticipated major rehabilitation costs for the existing structures. It is thus important to review how these rehabilitation cost savings were estimated.

The draft feasibility study report details two different methods used to assess rehabilitation costs for the "without project" condition. The first entailed an in-depth review of the costs and results of the current system rehabilitation program. Table 4.2 (from the draft feasibility study) shows the extent of this program. As shown, projects were carried out over the past 20 years at 36 sites, at a total cost of $905 million, an average of $25 million per site. Consistent with the current engineering policies of the three Corps District offices involved, it was concluded that additional rehabilitation projects would be needed at 25-year intervals to allow the

TABLE 4.2 Lock and Dam Rehabilitation Projects

Location	Date	Project Cost
UMR		(actual $'s in millions)
Lock and Dam 25	1994-2001	$25.9
Lock and Dam 24 (multiple stages)	1996-2007	$70.0
Lock and Dam 22	1987-1990	$15.1
Lock and Dam 21	1987-1990	$14.6
Lock and Dam 20	1986-1994	$43.7
Lock 19	FY04	---
Lock and Dam 18	1988-1993	$15.0
Lock and Dam 17	1988-1993	$14.9
Lock and Dam 16	1991-1994	$17.8
Lock and Dam 15	1993-1996	$19.2
Lock and Dam 14	1996-2000	$30.6
Lock and Dam 13	1993-1997	$22.5
Lock and Dam 12	2000-2004	
Lock and Dam 11	2001-2005	$40.3
Lock and Dam 10 (multiple stages)	1991-2005	$25.9
Lock and Dam 9 (multiple stages)	1993-2004	$28.2
Lock and Dam 8 (multiple stages)	1992-2002	$25.9
Lock and Dam 7 (multiple stages)	1993-2005	$32.2
Lock and Dam 6 (multiple stages)	1991-2002	$25.9
Lock and Dam 5A (multiple stages)	1992-2002	$24.4
Lock and Dam 5 (multiple stages)	1990-2001	$36.1
Lock and Dam 4 (multiple stages)	1989-2003	$30.0
Lock and Dam 3 (multiple stages)	1988-2003	$32.7
Lock and Dam 2 (multiple stages)	1987-2003	$32.5
Lock and Dam 1 (multiple stages)	1983-2003	$56.2
Lower St. Anthony Falls	- - -	- - -
Upper St. Anthony Falls	- - -	- - -
IWW		
Starved Rock (w/o miter gates)	1981-1984	$13.3
	1981-1984	$16.7
Dresden Island (w/o miter gates)		
Lockport (w/o lock walls)	1983-1987	$22.7
Brandon Road (w/o miter gates)	1984-1988	$23.7
Marseilles Dam	1985-1990	$15.7
Peoria Dam	1987-1991	$23.2
LaGrange Dam	1987-1991	$21.5
Lockport Lock (lock walls)	1993-1996	$13.0
Brandon Road (miter gates)	1993-1996	$6.5
Dresden Island (miter gates)	1993-1996	$6.0
Marseilles (miter gates)	1993-1996	$6.0
O'Brien Lock and Dam	FY03 Start	$19.3
TOTAL		$904.7 million *

* Note: Project costs and totals represent actual expenditures that are not adjusted for price levels.

SOURCE: USACE, 2000a.

existing structures to perform satisfactorily for the 50-year study period. The anticipated typical costs of a rehabilitation project on a lock or dam are listed in Table 4.3.

The specific years that rehabilitation would be needed were based upon consideration of the number of lockage cycles that had occurred prior to the current rehabilitation program, and the number of years that it would take to accumulate a similar number of lockage cycles under the projected traffic. This analysis is summarized in Table 4.4. A schedule of lock rehabilitation expenditures showing project costs of $25 million or $30 million at each site on two occasions during the 50 years was developed (see Table 4.5). For example, Lock 25 would require a $30 million rehabilitation in 2020 and a $25 million rehabilitation in 2045.

The second method of estimating rehabilitation needs was an application of probabilistic risk assessment, using event trees to capture potential failure modes and repair costs. In this analysis, the engineering work group focused on 18 components of the locks and dams and projected their reliability to see if replacement or rehabilitation was economically justified before those components failed and needed to be replaced. Using a 1993-1995 version of the risk assessment model, no rehabilitation work appeared to be economically feasible for any component, leading to a "fix as you go" policy that would not repair aging parts until they failed. The findings were not believable, given the historic record of needed repair and rehabilitation. For example, only $1,000 was estimated as being needed for the next 50 years of miter gate repairs for all sites—a drastic understatement of real needs, given the past history of repairs. The repair estimates also did not cover the costs of restoring paint at even one site for hydraulic steel structures over the entire planning period (USACE, 2000a). Hence, this method was abandoned in favor of the analysis of historical lock and dam rehabilitation expenditure data described above. While the probabilistic risk assessment failed for the present study, the approach seems to have promise, particularly for analyzing circumstances that might arise that are not covered adequately by a historical record. The committee encourages the Corps to pursue this topic further as a research project, with a strong emphasis on using historical data to inform the theoretical analysis. The aim of this research should be to develop better analytical risk assessment methods that can be combined with other, improved planning and economic models to insure that maintenance policies and rehabilitation projects, and related risk assessments, are given proper weight in future feasibility studies, as directed by the Corps' planning guidance.

For the "with project" condition, where new lock construction is involved, rehabilitation cost savings arise because most of the major rehabilitation items will be included as part of the new construction. For example, damaged concrete will be repaired, lock machinery and filling and emptying components will be replaced or modified, and lock gates will be replaced. Hence, the first cycle of major rehabilitation in the "without project" schedule can be deferred. The rehabilitation costs thereby avoided are legitimately counted as a project benefit. Rehabilitation projects will still be needed for the extended locks, but they will occur later than in the "without project" condition. In the case of Lock 25, for example, the first major rehabilitation would be scheduled for 2035, rather than 2020 (see Table 4.2).

The importance of rehabilitation cost savings caused the Corps to closely evaluate the timing of the lock chamber extension projects, under the "optimal timing" analysis reported in the draft feasibility study, to ensure that the proposed lock construction would coincide with the timing of the first rehabilitation cycle. Ideally, lock construction would be delayed until near the

TABLE 4.3 Future Rehabilitation Project Costs Per Site (FY 1997 Price Levels

Feature	Project Cost
Lock w/only localized concrete repairs	$25,000,000
Lock w/complete chamber concrete resurfacing	$30,000,000
Dam	$15,000,000

SOURCE: USACE, 2000a.

TABLE 4.4 Number of Cycles Versus Rehabilitation Years

Lock Site	Year of First Traffic	Year of First Rehab	# of Cycles	Year Cycles Repeat	# of Cycles	Rehab Cycle in Years	Year Cycles Repeat		# of Cycles
25	1940	1995	353,0	2022	357,00	27	2048	367,00	26
24	1940	1995	331,0	2021	334,00	26	2046	343,00	25
22	1940	1990	266,0	2013	269,00	23	2033	264,00	20
21	1940	1990	272,0	2013	269,00	23	2034	274,00	21
20	1940	1990	269,0	2014	273,00	24	2035	266,00	21
19	1940	2005	225,0	2042	225,00	37	---	---	---
18	1940	1995	305,0	2021	305,00	26	2046	305,00	25
17	1940	1995	289,0	2021	289,00	26	2046	289,00	25
16	1940	1995	312,0	2022	312,00	27	2048	312,00	26
15	1940	1995	408,0	2023	408,00	28	2050	312,00	27
14	1940	2000	439,0	2033	439,00	33	2065	439,00	32

Notes: Lock 19 is a 1,200-foot lock chamber; thus, number of cycles is lower for same traffic condition.

1940 was the first year of significant traffic on the system.

Repeat cycles based on whole year intervals.

SOURCE: USACE, 2000a.

TABLE 4.5: Scheduled Lock Rehabilitation Expenditures ($ Millions)

	Lock 25		Lock 24		Lock 22		Locks 20 & 21	
Year	Without Project	With Project	Without Project	With Project	Without Project	With Project	Without Project	With Project
2010	-	-	-	-	-	-	-	-
2015	-	-	-	-	30	-	30	-
2020	30	-	25	-	-	-	-	-
2025	-	-	-	-	-	-	-	-
2030	-	-	-	-	-	-	-	-
2035	-	25	-	-	-	25	-	-
2040	-	-	-	25	25	-	25	25
2045	25	-	30	-	-	-	-	-
2050	-	-	-	-	-	-	-	-
2055	-	-	-	-	-	-	-	-
2060	-	35	-	-	-	35	-	-
2065	-	-	-	35	30	-	30	35

TABLE 4.5 (continued)

	Locks 15 – 18		Lock 14		Peoria & Lagrange	
Year	Without Project	With Project	Without Project	With Project	Without Project	With Project
2010	-	-	-	-	-	-
2015	-	-	-	-	30	-
2020	30	-	-	-	-	-
2025	-	-	30	-	-	-
2030	-	-	-	-	-	-
2035	-	-	-	-	-	-
2040	-	25	-	25	25	25
2045	25	-	-	-	-	-
2050	-	-	25	-	-	-
2055	-	-	-	-	-	-
2060	-	-	-	-	-	-
2065	-	35	-	35	30	35

SOURCE: USACE, 2000a.

scheduled rehabilitation date, but not so long as to require rehabilitation to maintain lock efficiency and safety. Specifically, it was assumed that the without-project rehabilitation cannot be avoided if construction of lock improvements begins more than five years beyond the date of the scheduled rehabilitation. This analysis showed that delaying the extension projects to allow capture of the rehabilitation cost savings would have positive benefits in all cases. This has the further advantage of delaying the construction expenditure, thus reducing its discounted present value. A proposed delay in project timing is consistent with this report's recommendation to not consider irreversible construction commitments until nonstructural alternatives are first carefully considered—a measure that will allow time to determine if projected lock traffic increases are actually materializing.

The Corps conducted a thorough analysis in considering rehabilitation costs and benefits, and the draft feasibility study report adequately documents this effort. Avoided rehabilitation costs are a legitimate part of the estimated benefits of the lock extension projects. The feasibility report could be improved by showing explicitly that the project construction cost estimates do, in fact, include costs for the major rehabilitation that will be conducted as part of lock extension projects.

This committee's assessment comes with an important caveat. It is clear that rehabilitation cost savings are tied closely with the proposed timing of the rehabilitation projects, which, in turn, is tied to the number of lockage cycles projected to occur to serve forecast demand. This committee has been critical of the demand forecast. If new waterway traffic demand forecasts are developed, it will be important to revisit the rehabilitation costs analysis to ensure consistency with the revised traffic demand forecasts.

Construction Costs

Because construction costs are uncertain, the Corps includes a standard "contingency" factor of 20 percent to account for possible cost escalation. The percentage allocated for contingency in the draft navigation feasibility study has varied from 20 percent to 35 percent. Since the Corps chose a construction method not used previously on the UMR–IWW, the contingency was raised to 35 percent. Subsequent design modifications, which eliminated some of the previously proposed innovative design and construction techniques, led the Corps in July 2000 to conclude that a 25 percent contingency figure was justified. Also, by July 2000, the Corps felt it had become more knowledgeable about site conditions and that more information was available on the proposed construction methods, thereby supported the decision to reduce the contingency figure to 25 percent. Regarding site conditions, when the Corps revised the contingency figure, the Corps reported that it was able to analyze results from successful field tests and subsurface investigations at some specific sites. Furthermore, it was reported that the Corps had information (from hydraulic modeling of the sites at Locks and Dams 22 and 25 and from emptying and filling models obtained from the Corps' Great Lakes and Ohio River division) that further supported the decision to reduce the contingency factor to 25 percent.

Based on the information provided to this committee, a slightly higher contingency figure (25 percent, rather than 20 percent) may be warranted for the new construction method. However, a recent construction effort on the UMR–IWW—construction of a second lock at Lock and Dam 26—cost nearly three times the estimated cost. Clearly, a 20 percent contingency was too small in that case. In construction projects, many factors can raise costs far beyond the original estimate, as demonstrated at Lock and Dam 26. The committee judges the risks of cost escalation to be far greater than the 20 percent or 25 percent estimates that the Corps has been using. If lock extensions on the UMR–IWW are sensitive to construction costs being no more than 25 percent above the Corps' estimate, it is likely that escalating costs will greatly reduce the project's net benefits.

The Value of Independent Peer Review

Large and important projects such as proposed lock extensions on the UMR–IWW would benefit from a second opinion. Whether the issue is surgery, revising the Head Start program, or extending locks, issues such as these are too important to not receive an independent judgment on the merits of the various approaches and a careful scrutiny of the analysis. There are nearly always different possible approaches to achieving a goal; decision-makers need assurance that the analysis was conducted carefully and represents the state-of-the-art. Before embarking on important decisions—particularly ones involving more than $1 billion of construction—an independent peer review is crucial. The final feasibility study of the UMR–IWW should be given a thorough independent peer review by an interdisciplinary panel of experts from outside the Department of Defense. The committee notes that Section 216 of the Water Resources Development Act of 2000 (WRDA 2000) mandated a National Academies review of the Corps' procedures for independent peer review of feasibility reports. This upcoming study should help broaden and strengthen the Corps' peer review procedures, which ultimately should lead to better feasibility studies.

5

Improving Waterway System Planning

Like many large water resources project planning studies, the Upper Mississippi River–Illinois Waterway system navigation feasibility study represents a complicated challenge. Even with a large budget, decades to complete the study, and an environment free of the contending interest groups, the conclusions would contain a significant degree of uncertainty. The committee recognizes the inherent difficulty of the Corps' task and our inability to give precise answers to the questions that the Corps must answer.

A major difficulty is the conflict between long-term infrastructure investments and the rapidly changing nature of the U.S. economy. Since the Second World War, the role of U.S. agriculture has changed from feeding the world through massive surplus production to exporting into a highly competitive world market for food and fiber. Increasing world population has increased food demand, but the Green Revolution and other innovations in agronomy have increased food supply throughout the world. American agricultural exports compete with exports from Argentina, Brazil, and other food exporters, as well with increasing domestic production in importing nations. The dynamic nature of the competition and changes in technology make it difficult to accurately forecast future levels of American food exports.

Infrastructure improvements, such as larger locks on the Upper Mississippi River, are designed to expedite river traffic for the next century or more. In evaluating the social benefits of these improvements, the benefits 20–100 years into the future are important. Unfortunately, rapid changes in the U.S. economy and in other national economies mean that the future value of these navigation improvements is uncertain. Growing world population, growing affluence, and innovations in agriculture could lead to either much greater or much lower demand for U.S. grain exports. Extending large locks on the Upper Mississippi could be similar to the example of the construction of the huge airport in Gander, Newfoundland, which was built to serve propeller aircraft that needed to stop on transatlantic flights. Rapidly increasing traffic at the airport in the early 1950s induced the authority to make an expensive infrastructure investment just as jets were being introduced that would fly nonstop between Europe and the U.S., thereby avoiding Gander.

A rapidly changing economy makes the future benefits of infrastructure investment extremely uncertain; however, this cannot and should not be a reason to delay or stop investment.

If it was decided to not invest in lock extensions, traffic on the waterway would likely not grow much beyond current levels because of the costs associated with congestion. Thus, although uncertainty about the future value of infrastructure investment should not be a reason for taking no action, infrastructure investments such as the Gander Airport and improvements to navigation such as the Tennessee–Tombigbee waterway make it clear that ill-conceived investments can be costly.

Improving the efficiency of current use lessens congestion and pushes back the time when a decision must be made on infrastructure expansion. As the Gander Airport example shows, delaying a decision for a few years could show that the investment should be designed differently or even is unnecessary. As soon as the Boeing 707 demonstrated that it was cheaper, more reliable, and much preferred by passengers, the airlines scrambled to buy jets and shed their propeller aircraft from their transatlantic traffic. Had they waited a few years before making their expansion decision, Gander planners would have seen that airport traffic was about to decline precipitously.

There are costs, however, to waiting for more information, just as there are costs to building a project on the basis of current information. If traffic continues to increase, waiting to build a project means that users will have to bear the costs of increased congestion for years.

Society needs a prudent rule for deciding when there is sufficient information to stop waiting and to start building. Such a rule involves tradeoffs between the social losses from building when it is not warranted, and delaying construction beyond the points where new capacity is needed. The decision also requires an estimate of the likelihood that the additional capacity is actually needed on the basis of currently available information. The latter is a technical decision; this committee can lend advice to the Corps on the estimation. The former is a value judgment that Congress—not the Corps—must make.

Extra throughput could be squeezed out of the current locks by improving congestion management. Although scheduling tows to arrive at the locks is difficult, valuable steps can be taken. The level of traffic on the Upper Mississippi River is not uniform over the navigation season. Smoothing traffic would significantly reduce congestion and delays. Nonstructural measures such as a scheduling system, congestion pricing, and tradable permits will also initiate smaller and far fewer environmental impacts than structural measures such as lock extensions. They are therefore more consistent with strategies for the sustainable development (promoting improved traffic flow and environmental restoration) of the Upper Mississippi River and tributary system.

When infrastructure investments are prudent, it is important to find the appropriate scale for the investments. Social resources are wasted by building a small structure that must be replaced in a few years, or by building a structure whose capacity is not needed. Thus, a careful analysis of the range of possible future demands and ways of accommodating them is needed.

The nation has undergone a profound shift in how environmental resources are regarded and valued. For example, the prospect of removing four large federal hydroelectric power dams on the Snake River in the Pacific Northwest, in order to restore salmon runs and the environment, has been discussed at the highest policymaking levels; several other U.S. dams have already been removed. These changes are occurring because environmental consequences have in some instances turned out to be worse than expected, and because society today places greater value on not damaging ecological resources. The lesson is that infrastructure changes last for a long time, long enough for people's perceptions and preferences to change. In the committee's judgment, environmental concerns will be continually important in the future.

This chapter comments on these four analytical topics—making decisions under uncertainty, short-term nonstructural measures, longer-term measures, and environmental considerations and studies. It is provided to the Corps in the spirit of helping improve the draft navigation feasibility study.

CONTEMPORARY UNCERTAINTY ANALYSIS: THE STATE OF THE ART

Analysis Under Uncertainty

Complicating the Corps' analysis is the fact that infrastructure investments have long time horizons. About a decade would be required to extend the locks on the Upper Mississippi River–Illinois Waterway, after which they would be expected to last perhaps a century, with periodic maintenance. No one can know or predict with confidence the demand for water transport—or almost anything else—50 or more years in the future. In this section, we examine how to analyze uncertain future demands. Past Corps analyses did not recognize or treat uncertainty explicitly, even when dealing with shipments up to 50 years in the future. Failure to deal explicitly with uncertainty leads the unwary to have far too much confidence in the resulting forecasts and analysis, which can lead to bad public decisions concerning waterway investments.

Sensitivity Analysis

A common way to treat uncertainty is by conducting a "sensitivity analysis." As an example, demand for grain shipments on the Upper Mississippi could be explored along the entire range of what is plausible in each year up to 2050. In particular, grain could be sold to local mills for processing or exported via a route other than the Upper Mississippi waterway. If so, barge shipments of grain on the waterway could be reduced to zero. At the other extreme, increases in world population combined with increased affluence and a growing demand for meat could increase world prices for grain to the point where much larger tonnages move on the rivers. Such a sensitivity analysis would find that the benefits of extending the locks would be small if grain shipments fell and would be substantial if traffic increased greatly. This method can be helpful if the full range of future values for the crucial variables is specified. It is easy to bias the analysis by artificially truncating the range by, for example, assuming that grain shipments could not fall. However, when the full range of future values is specified for all the important variables, the outcome of an uncertainty analysis is somewhat predictable. In general, one set of values produces demands too small to justify expansion, while another set of values produces demands that make expansion a priority investment. Although the analysis can isolate the crucial variables and the values of these variables, other methods must be used to assess the likelihood that demand will be high enough to justify lock extensions.

Monte Carlo Analysis

A second method for treating uncertainty builds on sensitivity analysis by specifying a probability density function (distribution) for each uncertain variable. Unlike sensitivity analysis, which attempts to specify the plausible range, this second method—Monte Carlo simulation—attempts to place probability estimates on each value in the range. Once these distributions are specified, a Monte Carlo analysis is used to calculate the posterior distribution, given the structure and distribution of each variable. The Corps currently uses this method in its flood damage reduction studies (NRC, 2000).

If one has a good idea of the probability distributions for future values of each variable, and if those distributions can be specified with confidence, the resulting posterior distribution could be extremely informative. It would give the probability that the benefits would be high enough to justify extensions. Sensitivity analysis might conclude that some values of future demand are so low that lock extensions are not justified. In contrast, the Monte Carlo analysis might conclude, for example, that there is a 90 percent chance that lock extensions would be economically justified. This analysis would give the Corps and Congress information on the likelihood that lock extensions are justified—if there is general agreement on the probability distributions for the values of the crucial variables.

Unfortunately, analysts are hard-pressed to specify the range of uncertainty, much less the probability distribution. In these cases, analysts assume that the probability distribution is uniform or triangular or some other convenient, plausible, but unknown shape. In general, the posterior distribution and the probability that lock extensions are justified depend on both the range of uncertainty and the shape of the assumed distribution. Because the distribution is not known with certainty, the analysis should be conducted by examining all plausible distributions, a task rarely undertaken. The information that comes from a Monte Carlo (or any other type) analysis cannot be greater than the information and assumptions that went into the analysis. In other words, if one knows little or nothing, an analysis is not likely to provide a great deal of useful information. This obvious point, however, is often forgotten by analysts and by the sponsors of their analysis. If an analysis produces powerful results—despite the fact that little information was used as input to the analysis—these results should be viewed with some suspicion. In a Monte Carlo analysis, this might occur because the analysis unknowingly makes strong assumptions about the probability distributions.

Scenario Analysis

A third approach is the use of "conditional forecasts" or "scenario analysis." For example, one scenario would be expanding world demand for grain while another might be increasing competition from grain producers in other nations. Each of the scenarios is conditioned on (one assumes) a value for the crucial variables, and then explores the implications of lock extensions. In this sense, scenario analysis is a development of sensitivity analysis where particular values for each crucial variable are specified by sketching a particular future on the basis of some stated assumptions. The social gains and losses associated with lock extensions under each scenario can be explored. If probability values can be placed on each scenario, the analysis becomes similar to the Monte Carlo analysis described above. If not, each scenario becomes a way of thinking about the future. Scenario analysis is more conceptual and qualitative than it is quantitative. It can produce insights but it rarely produces useful estimates.

Wait and See

A fourth approach is to recognize that the parameters really are unknown and that it is not necessary to make a final decision today. In this approach, the social costs and benefits of making a decision today are constantly evaluated against the social costs and benefits of waiting for more information.

But the future can hold discontinuous changes, such as catastrophic failures of infrastructure. When infrastructure will be subjected to the possibility of extreme events, engineers usually aim to account for such possibilities in project design. Dams, for example, may be constructed to withstand a "probable maximum flood." However, decisions about appropriate design criteria, or when a project should be built, have little to do with "wait and see" analysis. Society should not allow crucial infrastructure capacity to lag behind demand and should not neglect inspection and repair on this infrastructure. Neither of these principles conflicts with a "wait and see" approach.

This approach becomes more attractive when steps can be taken to lower the cost of delaying the decision. For example, nonstructural improvements such as helper boats and improved scheduling, reduce traffic congestion at every level. Instituting such improvements means that the cost of delaying structural improvements is lowered, allowing more time to assess future demands for U.S. grain exports.

Finding Robust Strategies

Much of the future is unknown and unknowable. It would be wonderful if the Corps knew precisely how much barge traffic was going to use the UMR–IWW over the next 100 years. But they do not, and no amount of analysis can predict precisely the amount of future waterway traffic. Rather than designing the lock and dam system to minimize costs for a precise (and unknown) amount of future traffic, it would be prudent to seek a construction program that provides significant benefits for a range of future traffic levels, even if it is not optimal for any single amount of waterway traffic. Robust policies—those that produce favorable outcomes under the full range of plausible scenarios—should be sought.

This search begins with a scenario analysis to define the range of likely levels of future waterway traffic. The costs of a wide range of construction scenarios are then considered together with the congestion costs, and other social costs, for each traffic level for each construction program. The goal is to identify a construction program that produces reasonably high benefits or low costs over the range of plausible outcomes.

This approach can also be combined with a "wait and see" approach. For the range of plausible scenarios, when will the true nature of future demand be revealed? In view of the costs of building before there is adequate information on the one hand, or the costs of waiting too long on the other hand, which alternative produces high benefits and low costs?

Comparing Approaches

Each of these approaches can be helpful, depending on what is known about the future. If a single variable is crucial, a sensitivity analysis can isolate the critical value that determines if expansion is worthwhile. If so, subsequent analysis can focus on the likelihood that this critical value is likely to be exceeded. If more than one variable is important and the analyst has a good idea of the distribution of each variable, a Monte Carlo forecasting approach can prove helpful. Just knowing that a small number of values of one or a few crucial variables are important can allow a scenario analysis that helps clear away confusion. Finally, finding a robust strategy that is beneficial across the range of uncertainties may require little information about the future. The best approach to each situation depends on the nature of the situation and what the analyst knows.

Generally speaking, some things are known about the future. A good deal is known about what the next few years will bring, but less is known as one looks further into the future. A major problem in uncertainty analysis is that analysts tend to be overconfident about their knowledge of the future. In many situations, actual events turn out to be outside the range of forecast futures.

All the options considered seriously by the Corps in formulating plans for navigation enhancement appear to involve large-scale lock and dam extensions and would require large public expenditures. A benefit-cost analysis to evaluate these expenditures requires information about waterway traffic over the next century; no one can predict future waterway traffic levels precisely. If demand for barge-transport capacity increases rapidly in the future, expansion will be in the public interest. But even current levels of waterway traffic generate enough congestion and delays so that some response is in the public interest: if the demand for barge transport continues as it has been for the past five years, no extensions will be needed.

Fortunately, several "management" improvements in tow operations could produce benefits at little cost: industry "self-help" efforts, incentive/disincentive fee structures, scheduling changes, speed limits, and installation of better barge coupling/decoupling equipment. Indeed, these nonstructural improvements not only *can* be made, but they *should* be required before determining whether and when construction is needed. The benefits and costs of nonstructural traffic management plans should first be carefully estimated before estimating the benefits and costs of lock extensions.

SHORT-TERM (NONSTRUCTURAL) MEASURES

Congestion Problems on the Upper Mississippi

On the Upper Mississippi River–Illinois Waterway, towboats arrive randomly at locks.

This is expressed through unpredictable and random, sometimes very intense, periods of congestion. There is no perceptible pattern to arrivals of towboats at locks throughout the day. This leads to queues at locks of unpredictable lengths. Thus, for example, a tow that arrived at Lock 25 on October 8, 1999, at 11:35 p.m. locked straight through. A tow arriving one day later waited in a queue of more than 22 hours. A tow arriving two days after that locked through with no waiting.

Motorists are familiar with predictable traffic congestion patterns and plan their routes and departure times accordingly. Although drivers cannot completely eliminate congestion problems, they can work around them by changing routes and departure times. By contrast, imagine that traffic congestion was frequent and random and that one could not predict when or where traffic congestion would occur and could not alter one's route. If one wished to be sure to arrive at the office by 8 a.m. every morning, it would be necessary to leave hours early each day in case of an hours-long traffic jam that could not be avoided. Although one would almost always arrive by 8 a.m., on many mornings one would arrive at the office by 7 a.m., and much of the time it would be closer to 6 a.m., depending on traffic levels. Clearly this would involve a substantial waste of time and resources.

This is the position that inland waterway operators find themselves. *Through acting individually with no coordination, they have opted to treat lock congestion as an uncontrollable event.* Congestion slows movement on the system and results in millions of dollars of wasted resources. This raises shipping costs on the UMR–IWW and raises the rates that operators charge for their services. Because farmers' incomes are determined by the difference between the world price of crops and the cost of delivering crops to market, lock congestion on the UMR–IWW reduces incomes of the agricultural community in the Upper Midwest. Service quality is diminished because operators cannot give a reliable delivery date and time, thus diverting traffic to other modes. The key to solving the congestion problem on the UMR–IWW is to reduce the randomness of arrivals at each lock. If arrivals at locks became more regular and constant throughout the shipping season, system capacity could be increased, traffic could move more quickly, and rates charged to agricultural interests could be reduced.

For example, large airports limit the number of flights that can be scheduled to take off or land during an hour. But in the 1980s, few airports used this type of scheduling, which resulted in long queues to take off and many flights in holding patterns waiting to land. When airlines and passengers realized the frustration, high costs, and danger of such an unscheduled system, they agreed to adopt scheduling. The current system naturally works best under perfect weather and other ideal conditions, but it also generally works well when affected by flights delayed by maintenance, bad weather that reduces airport capacity, and other unexpected difficulties. Many airlines would like to schedule an extra flight during the period of highest demand. They understand that this additional flight will face some delays because of congestion. However, this additional flight would also impose costs (in the form of additional congestion and delays) on other users. These types of external costs caused by congestion must be internalized to have the system operate efficiently. For example, if an airline had to compensate other users for the additional waiting time imposed on them by this additional flight during a high-use period, these costs would discourage the airline from adding the extra flight.

An airport takeoff or landing "slot" at a large airport during a busy time is a valuable property. The federal government has essentially given airlines a property right to slots on the basis of when they first started scheduling flights in a given time window. Allowing airlines to buy and sell slots has enabled them to transfer slots to more valuable times and has given them more flexibility in planning their operations.

Airlines, passengers, airports, and the federal government find it hard to believe that they tolerated the congestion that existed without the current scheduling system. The current system has enabled airlines to schedule their crews more predictably and efficiently, has reduced their costs, serves passengers better, has reduced frustrating congestion delays, and has increased safety. No one is interested in returning to the old system.

Precisely the same principles apply to tows seeking to transit a lock. Congestion delays are frustrating and expensive and increase shipping costs. While each tow operator would prefer to have first priority at each lock, they understand that this is not possible. Rather than waiting many hours in a queue, tow operators would prefer to know when locks will be free so that they can schedule their fleets to optimize equipment usage. For example, if an operator knew that there was a window of clear sailing next week from St. Louis northward to Keokuk for pickup, then back downstream after pickup, the towboat could be dispatched at that time, thereby allowing productive use of the equipment on the Lower Mississippi until the entry time into the lock system. This increase in the hours per year during which the towboat was productively deployed would allow towboat operators to amortize their equipment costs over a larger traffic base and could reduce commercial cargo shipping rates on the UMR–IWW. Unless the introduction of traffic management schemes changes the market structure of the barge industry—a change not likely to occur, based on both theoretical arguments and historical experiences—the reduced costs will be at least partially passed on to the consumer.

There is another advantage to greater predictability of arrival times. If operators knew that there were time frames within which clear sailing through all locks would be possible, priority services could be developed for waterway operation. The development of highly dependable, high-speed services for high-value freight was one of the unforeseen consequences of the deregulation of railroad and highway motor freight. Although waterway freight is typically low-value and is thus not as attractive to priority services, the possibility that there is a demand for highly predictable, higher-speed waterway services—a demand currently not satisfied because of unpredictable congestion delays at locks—cannot be ruled out.

Congestion has been analyzed in many systems, from highways to ports to supermarket checkout counters. Some standard ways of handling congestion are (1) charge a congestion toll equal to the social cost imposed by each additional user, (2) require that users make reservations for using the system to eliminate excess demand, and (3) ration the number of users by selling or giving away slots that enable individuals to access the system at a particular time (such as takeoff and landing slots at Washington's Reagan National airport).

Each of these approaches can increase the efficiency of the system and lower the social cost. Market-based systems promise greater efficiency than approaches such as "first-come, first-served." Three issues are involved: (1) reducing waiting time at locks by better scheduling, (2) getting the tows with the greatest need for haste through first, and (3) minimizing the time each tow ties up the lock. A market system can provide incentives for improving all three aspects of efficiency. In contrast, a nonmarket system, such as first-come, first-served, can deal with only the first issue. Allowing tows to trade slots can deal with the first two issues. To address the third issue, tows must face an incentive to minimize the time in the lock, such as through a charge for each minute spent in the lock.

One market-based system is a congestion toll, where each tow pays for the cost of the delays imposed on other tows. This system deals with all three issues, but it can be complicated and is likely to result in high tolls at congested locks. A second market-based system involves issuing a property right for the amount of lockage time used in a historical period (see Appendix C for a discussion of this system). The lockage time could be broken down into 5-minute increments. By trading the increments, a tow operator could assemble the total time required to get a tow through a lock at the time when the tow needs to get through. Because the time is in 5-minute increments, each tow operator is motivated to minimize the time spent in the lock.

Each of these market-based systems is more complicated than the current system. Towboat operators, lock masters, and other current users of the system may react negatively to proposals for such large changes in the system for managing lock operation. However, market-based systems have proven to be highly beneficial in many applications. And, in the committee's judgment, the Corps and tow operators could quickly learn to use these systems and benefit from their implementation.

The primary advantage of a system of tradable slots is that it would allow towboat operators to plan their season's sailing schedule in a way that leads naturally to a smoothing of arrivals. This would speed the flow of traffic, allowing more trips for each tow per year, and would save on towboat costs, resulting in lower rates to users. There are other advantages as well.

The second most important advantage of the system is that it encourages efficient use of the locks themselves. There is a huge variation in the speed with which tows traverse locks. For example, in October 1999 at Lock 25, the average time for a tow requiring a double lockage to transit a lock heading downstream was 78 minutes. 95 percent of the tows took 114 minutes or less, while 5 percent of the tows took 48 minutes or less. For tows heading upstream that required a double lockage, the average time was 92 minutes. 95 percent of the tows took 127 minutes or less, while 5 percent took 64 minutes or less. Thus, the fastest tows took half as long as the slowest tows. If all tows could be made as efficient as the fastest current 5 percent, many more tows could be served by each lock, essentially eliminating the current congestion.

A system of lockage permits would give an incentive for inefficient operators to increase the speed at which they cleared the locks. Those operators who were slow to clear locks would find that they needed to reserve more lockage time than the faster operators. The increased costs would come from an operator's bottom line because users would choose the shipping company with the cheaper rates. Operators would respond, for example, by increasing the training of deckhands to speed recoupling, by adopting new coupling techniques or, when appropriate, by paying another operator for assistance in clearing locks. It must be expected that operators would experiment with methods than are now not currently used to see if there would be savings in lockage times. Of course, if average lockage times are reduced, more tows per day can use each lock. It is impossible to predict the degree to which new technologies explored in the effort to economize on lockage time would expand the economic capacity of the current locks. If history is a guide, however, the saving would be substantial. Providing clear title with trading privileges is one of a class of incentive-based policy instruments that could be used to induce more efficient use of navigation infrastructure. For example, incentive-based instruments have been adopted to address lead in gasoline, ozone-depleting chemicals, nitrogen oxide and sulfur emissions, new-vehicle fuel efficiency, urban land development, and retirement of older, heavily polluting vehicles. Emissions constraints and tradable permits have been proposed by the U.S. as a strategy for controlling global emissions of greenhouse gases. Incentive-based policy instruments include user fees, congestion pricing, pollution taxes, cap-and-trade policies, deposit-refund programs, and replacing subsidized prices for inputs and outputs with market-driven prices. Appendix C describes experiences with tradable permits in several recent applications outside of inland navigation.

These nonstructural options for improving waterway traffic management hold great promise for helping alleviate waterway congestion quickly, at relatively low costs, and in a manner that is more consistent with the long-term environmental sustainability of the Upper Mississippi River ecosystem than large-scale, structural changes. There are benefits and costs associated with each nonstructural option, the careful assessment of which was beyond this committee's scope and resources. As a first step toward improving UMR–IWW traffic management, a careful study of the benefits and costs of nonstructural options for improving waterway traffic management should be conducted.

Other Strategies for Alleviating Congestion

An alternative to offering transferable titles to lockage slots would be to allow an operator who was within three days of arriving at the first lock in a sequence to reserve lockage times at that lock and all succeeding locks to be traversed. This would increase the predictability of the lockages at succeeding locks. Any technique that increases the predictability of arrival times would help to eliminate wasteful development of queues at locks. We consider this system to be less desirable than a system of clear title to lockage times for several reasons:

- It would be necessary to determine how long a lockage time should be. As noted above, some operators have longer times and some shorter. If lockage slots were determined by the longest time, the locks would stand idle following a fast tow, thus wasting resources. If the lockage time was determined by the fastest operator, it would be necessary to develop a penalty scheme to penalize slow performance.
- Only limited smoothing of arrivals would take place. For example, an operator would have no assurance that he or she would not encounter a queue at the first lock. The operator would have no assurance that the desirable sequence of lockage slots was available upstream (for northbound movements, for example). Similarly, there would be less incentive for seasonal smoothing of demand.
- An operator would have an incentive to over-reserve slots. For example, assume that a towboat operator planned to travel upstream from St. Louis to Quincy, Illinois. As the operator approached St. Louis, the operator would be tempted to reserve lockage slots all the way to Lock and Dam 1 at Minneapolis, in the hope of having a smooth trip all the way up the river. In order

to avoid such gaming of the system, a program of penalties for reserving but not using lockage slots would be needed. This complication would not occur if all operators were given clear title to lockage times at the beginning of the navigation season based on historical usage patterns.

- Unlike the permitting scheme in which the bidding process for Corps-owned lockage slots provides a tangible and indisputable measure of willingness to pay (and thus a measure of the benefits of lock extensions), the value of lockage times would not be revealed by such a system. Users would thus be in the same uncomfortable position they are currently in—an inability to demonstrate, beyond a reasonable doubt, the advantages of lock extensions.

LONGER-TERM MEASURES

Traffic Demand Forecasts

Several Corps of Engineers papers have advocated the use of spatial equilibrium models to evaluate the economics of public investments. A 1998 report from the Corps' Economics Work Group, for example, states, "The consensus of this literature is that the economic impacts of transportation systems are best analyzed as components of larger spatial price economic models. Analyzing transportation systems or their individual components myopically can lead to erroneous conclusions regarding economic impacts and values of the transportation system and its components. Spatial price economic models may be characterized as models where consumer demands and producer's supplies of goods and services are identified by their location in spatially geographic regions called markets" (USACE, 1998). The committee agrees with this basic approach.

The above statements imply that an evaluation of investment in inland waterway navigation improvements should be based on disaggregated spatial equilibrium models where all relevant alternative supply and demand regions are identified and connected by product prices, alternative modes, and transport rates. Development of the disaggregated spatial equilibrium grain models involves five steps:

1. The first step is to forecast world import demands and U.S. grain export demands, as nearly all grain shipments on the Upper Mississippi River are destined for export markets. This requires forecasting both the demand for imported grains and the supply of exported grains. The former depends on population growth, per capita income, and the demand for meat. The latter depends on grain production in Argentina, Brazil, the U.S., and other grain-producing countries.
2. The second step is to forecast the amount of grain that each farmer will want to send to each market, e.g., local animal feed, processing, and exports. The expected grain price in each market, together with the shipping cost, will determine in which market and by which transportation mode the farmer can expect to get the greatest revenue.
3. The third step is to aggregate the net revenue maximizing decisions of individual farmers in order to calculate the market equilibrium for both the uses of the grain and the shipping modes. This is because the price for grain in each market will fall as farmers desire to ship more grain to this market, and the price of transportation will rise as congestion increases because of increased shipments.

4. The fourth step is to forecast the performance of each mode and route that is affected by the traffic volume carried on that link. The objective is to determine how the level of traffic affects congestion levels experienced by each link, and thus how rates charged by transport operators on those links change with traffic levels.

5. The final step is to estimate the net benefits of transportation infrastructure investments.

These steps require the development of forecasting and optimization models.

Forecasting World Import and U.S. Export Demands

There are two basic types of forecasting models: structural models and nonstructural, or time series, models.

Structural models attempt to explain the behavior of dependent variables as a function of one or more causal independent variables or equations. The two basic large-scale agricultural-sector structural models are the U.S. Department of Agriculture (USDA) agricultural-sector model and the Food and Agricultural Policy Research Institute (FAPRI) model. Both models were designed to make projections rather than forecasts; that is, they were designed to evaluate alternative agricultural policies. Their focus is to measure the difference between alternative policy options rather than to forecast absolute levels of specific outcomes. To be used in forecasting, these models must be reformulated by replacing the largely linear regression equations with more modern forecasting equations like exponential smoothing and moving average equations. These latter equations place greater weight on recent data. In contrast, regression equations weigh 10- to 20-year-old data as equal in value to recent data.

Nonstructural, or time series, models do not attempt to use theoretical relationships such as supply and demand. Instead, they treat the variable of interest as an element to be forecast from past data. Several types of time series models are available. These include Box-Jenkins, ARIMA (autoregressive integrated moving average), and ECM (error correction models). These models typically consist of a time trend, a seasonal factor, a cyclical element, and an error term. Both types of models are likely to be superior to a linear regression time trend. It is difficult to incorporate nonquantifiable variables into all quantitative models. Therefore, judgment must be used to adjust these models to recognize nonquantifiable information. Both approaches can be useful in forecasting exports a few years into the future. Time series models are most accurate in extremely short-term forecasting, while structural models are likely to provide better forecasts of conditions beyond about two years into the future. Unfortunately, forecasting 20–100 years into the future is beyond the capability of these models. Instead, the forecasts depend on population growth, increases in per capita income, the demand for meat, and advances in agricultural productivity, all of which are inherently difficult to predict.

Forecasting Model Accuracy

There are several methods of measuring model accuracy. One of the simplest ways is to compare model forecasts with actual values. Models can be used to forecast recent year outcomes, and these forecasts are then compared with actual outcomes. The actual values can be regressed on the forecast values and R^2 (R^2 measures the proportion of the variation in the dependent variable, e.g., shipments, accounted for by the regression) is used to measure accuracy. The higher the value of R^2, the more accurate the model.

Dealing with Uncertainty

There are large uncertainties in making long-range export forecasts. One method of obtaining the "best" forecast is to use several forecasting models. Then, after comparing forecasted values with actual values, the 2 or 3 highest R^2 "most believable" models can be used to calculate a "combined" forecast, formed as a weighted average of the 2 or 3 selected forecasts. The weights can be obtained from the regression of the actual values on the forecasted values. After selecting the supply origins—most likely individual countries or groups of counties—the quantities of grain available for shipment can be estimated by subtracting the amount of grain fed to livestock from the production in each supply region. Several types of models can be used to forecast the quantity of grain to move to each market by mode and time period. Among the simpler methods are the linear programming (LP) and quadratic programming models.

Data Requirements

Sweeney and Marmorstein (1998) suggest that the reason the ESSENCE model failed to include grain origins and quantities, alternative markets, product prices, and transportation rates from all origins to all destinations was because of a lack of adequate data. This section outlines the data necessary to run a disaggregated spatial equilibrium model for grain.

Origin Data. The farmer makes the basic decisions that determine grain flows. A farmer's decisions on where to deliver grain from the farm initiate the grain distribution process. However, modeling every farm within the relevant river basin study area would be very expensive. A more practical approach is to define the origins as one point in each county or group of counties. The quantities of grain available from each defined origin can then be estimated.

Transportation Rates. Grain sales and shipment decisions are made on the basis of commodity prices and current transportation rates, not on long-run railroad, barge, and truck marginal costs. Characterizing current farmer decisions and shipments to market requires knowing the prices that farmers have to select from. Although long-run marginal costs may be a rough approximation for truck rates, and a less good approximation for barge and rail rates, actual rates are needed to explain current farmer decisions and to estimate price sensitivities.

Three caveats are in order when collecting data. First, models should allow for barge and truck shipping rate seasonality by including several—preferably 12—time periods. Barge and truck shipping rates are constantly changing. Thus, attempts to remove "abnormality" from the rates by eliminating variability—are improper. In some supply regions, rail shipping rates are frequently lower than barge shipping rates. Secondly, barge, truck, and rail rate data should be for the same time periods during which the quantity data are collected. Finally, rates should be for specific days, rather than averaged over long time periods.

Quantities Shipped from Each Origin. Quantities available to be shipped from each origin—a county or group of counties—should be calculated by subtracting the amount of grain fed to livestock or otherwise used within the origin area.

Destination Data

Alternative Markets. A spatial equilibrium model must identify and incorporate all relevant markets. These markets can be identified in several ways. The best method is to conduct grain flow surveys in each state. These survey data are useful not only for identifying relevant markets, but also for calibrating modal grain flows and modal shares. In addition to asking barge terminal operators, another option to identify relevant markets that should be included in the model is to ask the grain shippers who make the sales decisions.

Market Prices. Market prices are just as important as transportation rates in determining grain flows among the competing markets. These bid prices, for days that must exactly match the dates for the rates and quantities, can be obtained from grain buyers.

Market Constraints. Most domestic grain markets face quantity constraints for given time periods. These constraints are based on daily processing and storage capacities. Animal feeding markets face both minimum and maximum quantity constraints. The maximum constraint is based on the number of animals fed, and the minimum constraint is based on the fact that the market must have grain supplies each day or the animals will starve. Data on processing market constraints can be obtained from newspaper articles and by asking industry people about processing capacities.

There should be no constraint on export markets other than the export projections. However, there may be environmental or governmental policy constraints on total imports into individual countries. These constraints must be built into the export projections.

Developing a Disaggregated Spatial Equilibrium Model for Grain Transport

A spatial equilibrium model seeks to find the most profitable destination and transport mode for grain from each area. These models recognize explicitly that grain is grown in tens of thousands of locations and that there are hundreds of markets for the grain in the form of individual feedlots, processors, and export locations. Such models are most accurate when operated on a detailed spatial scale, such as a few farms or a county. These models summarize the historical relationships between the quantity of grain shipped by barge and barge shipping rates, the rates for other modes, and the prices offered for grain in various domestic and export markets. A model developed on historical data can be used to forecast future barge shipments on the basis of assumptions about the future rates for each transport mode, the amount of grain grown in each area, and the quantities of grain that are used as feed, that are processed, or that are exported.

All participants in the grain distribution system are assumed to seek maximum profits. Therefore, the spatial equilibrium model—a programming, network or other disaggregated optimization algorithm—seeks the shipping routes and modes that maximize total profits of farmers and transporters. The following steps should be used in developing a disaggregated spatial equilibrium model for grain:

Step 1: Identify the relevant geographic grain production areas to be included in the model. This should include all areas that ship grain on the Upper Mississippi River at least once a decade. These geographic areas should be subdivided into the largest number of (grain) origin areas that can be handled easily by the program (down to individual farms, if possible). The areas should be no larger than a county. One location within each area should be identified as the representative shipping origin for all grain to be shipped out of the area. The quantity of grain to be shipped out of the area should be defined as production minus the amount consumed within the area as animal feed or processed in mills; alternatively, within-area uses could be included as competing demands for the grain).

Step 2: Collect delivered grain prices for all destination markets to which this grain might realistically be shipped. These prices should be collected for one selected day for each month in the year(s) to be included in the analysis.

Step 3: Collect transportation rates for each relevant mode of transport from each shipping origin to each destination market for the same days on which delivered prices were collected. These transportation rates and grain prices would be used to calibrate the model to replicate actual grain flows for the year(s) from which prices were selected. In some studies, long run marginal costs (LRMC) for barges, railroads, and trucks have been used as surrogates for actual shipping rates. Unfortunately, actual shipping rates can be much more or less than LRMC at particular times because of circumstances such as high or low demand, heavy congestion, or weather-related difficulties. A model based on LRMC would fail to explain the drop in barge traffic when shipping rates were much greater than LRMC, and vice versa.

The committee strongly recommends that the Corps obtain and use actual shipping rates—not published tariffs—and corresponding quantities shipped for each model.

Step 4: Use the programming, network, or other optimization model to find the grain flows and transport modes that maximize the system profits—profits to farmers, shippers, and carriers—in shipping to domestic users and U.S. export ports. Market prices and transport rates should be used to calibrate the model based on past grain flows. To validate model results, the grain flow forecasts from the model should be compared with historical data (ideally, data not used to calibrate the model). The model should then be adjusted to correct for major discrepancies between the model and survey results. Changes in barge and other transport rates and in grain prices for various uses and destination can be used to estimate how sensitive the demand for barge transport is to various prices.

Step 5: Forecasting future benefits. While forecasting traffic flows depends on forecasting freight charges, the estimation of future benefits depends on long-run marginal social costs (LRMSC), which are defined as private (costs to the carrier) plus public costs. In forecasting future benefits, long-run marginal social costs should be forecast for each mode. The price for each mode will be long-run marginal social costs, plus or minus a factor that depends on the level of demand. The relationship between long-run marginal social costs and freight tariffs can be estimated with data from the past decade or so. The same relationship between long-run marginal social costs and shipping rates can be assumed to hold for the future. Public costs for barges include annual government expenditures on maintenance and operations, and Corps of Engineers expenditures for administration and research. Public expenditures for railroads should include annual government expenditures for research and labor retirement and federal, state, and local subsidies for short line railroads. Public costs for trucks include damage to highways in excess of fuel and use tax collections, and research expenditures. These and other such costs should be included in long-run marginal social/transport costs.

Step 6: The long-run marginal private and public modal costs should be adjusted to reflect the wide swings in transportation rates within and between years. Rate data should be collected in past years of high and low transport demand. These rate data from the past give perhaps the best indication of whether actual shipping rates will lie above or below long-run marginal social costs in the future.

Step 7: The effects of nonstructural improvements on the cost of barge freight should be incorporated into the long-run marginal private and social barge costs. Then the spatial equilibrium model should be run to estimate a base solution with which to compare the cost savings of structural waterway investments.

Step 8: Using the traditional Corps method of comparing the benefits of waterway grain transportation, run the base solution model and the after investment model to estimate the benefits of waterway investments under alternative scenarios of high and low demand and high and low transport rates—which are calculated from the long-run private and public cost estimates.

A spatial equilibrium model that relates actual traffic flows on each mode to the relevant shipping prices and local demands would be a large contribution to the Corps, farmers, and exporters. Not only would it specify the relationships between relative shipping rates for each mode and its level of traffic, it would specify traffic flows, document market volume, and provide myriad useful results such as: quantities shipped from each origin to each market, by mode(s); changes in shipments by origins; changes in receipts by destination; and shipments by mode(s) of transport. The price sensitivity of the quantity of grain shipped by barge could also be estimated from these model results.

ENVIRONMENTAL ANALYSIS

Adaptive Management

The emerging concept and practices of adaptive management are gaining currency within natural resource management programs and with many scientists and scholars. The principles of adaptive management grew from perceived shortcomings of traditional resource management practices, especially in large, complex ecosystems characterized by a high degree of uncertainty and occasional "surprises," such as floods or fires.

Many of the adaptive management concepts being applied and refined today can be traced back to research initiatives in the 1970s at the International Institute for Systems Analysis (IIASA) in Laxenburg, Austria and at the Resource Ecology Center at the University of British Columbia. A widely-cited document that resulted from the IIASA studies is *Adaptive Environmental Assessment and Management* (Holling, 1978). That volume identifies several key principles of adaptive management and emphasizes the interplay of ecological and social systems in enhanced understanding and improved ecosystem management. It also emphasizes the importance of the flexibility of resource management policies, and it promotes an iterative process in which policies are continually updated as knowledge of environmental behavior is gained. As Holling states, "the process of adaptive management and policy design . . . integrates environmental with economic and social understanding at the very beginning of the design process, in a sequence of steps during the design phase and after implementation" (Holling, 1978). Some of the components of an adaptive management approach include: (1) a shift from "trial-and-error" management to formal experimentation, recognizing that all management policies are experimental in nature, (2) careful monitoring of ecological and social effects and responses to management actions, (3) an integrated scientific approach to resources management that encompasses ecological, social, and economic considerations, (4) the investigation of uncertainties through formal experimental design and hypothesis testing, and (5) a commitment to ongoing adjustments in resource management based on the results of prior experiments (see also NRC, 1999b).

Given the topic's complexity and richness, specific definitions and practices of adaptive management continue to be refined. Another relevant definition of adaptive management is "a systematic process for continually improving management practices and policies by learning from the outcomes of operational programs" (Nyberg, 1998, p. 2). The concept continues to attract great interest, as evidenced by the adaptive management approach to resource management being applied by the U.S. government in the Pacific Northwest (Volkman and McConnaha, 1993), in the Florida Everglades, and in the Colorado River below Glen Canyon Dam (NRC, 1999b).

Adaptive Management on the UMR–IWW System

The UMR–IWW presents a unique challenge and an opportunity in which to apply adaptive management principles described by, among others, Holling (1978), Walters (1986), Lee (1993), and Gunderson et al. (1995). There have been no formal efforts at merging adaptive management principles with navigation system operations on the Upper Mississippi or on any other of the nation's navigation systems. Nonetheless, several elements of the UMR–IWW and the navigation feasibility study suggest that an adaptive management framework could help provide unity and cohesion to the feasibility study.

One principle of adaptive management is to avoid premature foreclosure of management options in order to avoid the trap of irreversibility: "We cannot always require a complete return to starting conditions or complete freedom to reach any other conceivable situation. But we can try to keep from getting locked into any one situation. No guarantees exist, but to ask honestly what options are being foreclosed reorients the planning and development process and makes dead ends less likely" (Holling, 1978).

In the Corps' feasibility study, the extension of locks is contemplated. Conducting an assessment of the benefits and costs of nonstructural options before considering lock extensions would provide the Corps additional time to study and learn more about the environmental effects of society's activities on the Mississippi River and throughout the river system.

An adaptive management approach would place environmental and social considerations on par with economic considerations. As Holling stated, "Environmental dimensions should be introduced at the very beginning of the . . . policy process design, and should be integrated as equal partners with economic and social considerations" (Holling, 1978). Given the links between ecology, society, and economics in the Upper Mississippi region, an adaptive management strategy would aim to balance these considerations at the outset of an investigation regarding an expanded navigation system. Encompassing these different factors within a multidisciplinary decision making paradigm is also consistent with the notion of sustainable development. In a more recent volume on adaptive management (Gunderson et al., 1995), Holling states: "Sustainable development is neither an ecological problem, a social problem, nor an economic problem. It is an integrated combination of all three" (Holling, 1995).

Adaptive management entails careful environmental monitoring and assessment in order to determine how management policies affect environmental systems. In the UMR–IWW system, further analysis of how the navigation system has affected river ecology would be consistent with an adaptive management strategy. Within the EMP for the Upper Mississippi River, the Corps and its cooperating federal agencies continue to enhance knowledge of how the navigation system has affected river ecology.

Adaptive management does not avert the challenge of forging a workable vision for the desired state of the ecosystem, and balancing between environmental quality and human uses of resources. Adaptive management ideally begins with a set of management objectives and then experiments to observe how those objectives may be better and worse served by various practices. The absence of well-defined management objectives is a common criticism of adaptive management efforts. This criticism would also haunt adaptive management efforts on the UMR–IWW if clear environmental quality and navigation use objectives are not established. Effective adaptive management for the UMR–IWW system would require a sustained effort to define management objectives (including input from the public and formally incorporating public opinion into management objectives) and to identify which ecosystem indicators and which human uses of the waterways will be included in adaptive management experiments. Through numerous public meetings in which it discussed the feasibility study, the Corps has laid the groundwork for meaningful public involvement.

Adaptive management is a long-term process and can be costly—both in terms of monitoring and research and in terms of the opportunity costs of deferring management changes until such changes are evaluated by adaptive management experiments. If adaptive management is to be effectively implemented in the Upper Mississippi River, it will require a sustained level of long-term funding for ecosystem monitoring and evaluation.

The resilience of the politics and institutions governing the resources at stake is another important consideration. Are these governing structures likely to allow changes in resource management in response to a new understanding of science-policy relations based on results from ecosystem monitoring and evaluation?

Independent review of scientific results and their inferences for policy innovation is essential to the integrity of ongoing adaptive management programs. Without such independent review, adaptive management processes can be circumscribed by stakeholder concerns, and the most essential scientific questions may be set aside because of political pressures. Independent review of scientific findings is also valuable when policy recommendations for promoting adaptive management are unpopular with some stakeholders. Several review "teams" are already in place on the UMR–IWW feasibility study, although the committee recommends that external, expert review—independent of potential conflicts of interest—be conducted as part of the UMR–IWW feasibility study.

Given the potential for an adaptive management strategy to help manage the multiple resources in the Upper Mississippi River Basin, the Corps should pursue the navigation feasibility study within the principles of an adaptive management framework. The Corps should seek the authority and funding from Congress necessary to conduct the study within an adaptive management framework.

The Corps' Adaptive Mitigation Strategy

Absent a fully implemented adaptive management approach, the Corps describes its approach to mitigating the environmental effects of incremental increases in waterway traffic as "adaptive mitigation." Although the introduction to the Corps' adaptive mitigation strategy demonstrates confusion over the respective definitions of adaptive mitigation and adaptive management, with these terms being used interchangeably (USACE, 2000b; preliminary draft EIS, pp 6–5), the Corps recognizes the value of adaptive approaches. However, a strategy that merely seeks to mitigate future negative environmental effects—without some type of environmental assessments and feedback during the expansion of the navigation system—does not reflect the principles of adaptive management as described in the natural resources management literature. For example, the feasibility study treats environmental issues and resources as planning constraints that require mitigation activities, rather than as valued resources that are to be protected or enhanced. The practical steps outlined in the Corps' adaptive mitigation strategy are generally reasonable; however, the committee rejects the notion that environmental resources should be treated only as constraints in the proposed expansion of the navigation system.

Improving the UMR—IWW Environmental Analyses

A first step toward improving environmental studies in the UMR–IWW feasibility study would be to take actions to address criticisms from local, state, and federal environmental interests. When considering this report's recommendations for improving environmental analysis, it should be recognized that the feasibility study was bound by earlier decisions that limited analysis to incremental effects and focused analyses to site-related project impacts. Furthermore, the feasibility study is being conducted in the context of a history of environmental investigations that includes ecological research on the UMR–IWW system, assessments of impacts from the second lock at Lock and Dam 26, and Environmental Management Program data gathering efforts. Finally, time is also important when considering both the sequence of project activities and the overall project planning period of 50 years. Sequence is important because protection of both ecological resources and navigation capability in the Upper Mississippi River basin depends on an adequate knowledge base for ecological impact analysis and for construction. If more is known about ecology, that understanding can be used with knowledge of actual construction practices and construction activity timing to avoid or mitigate impact.

Important long-term planning considerations also include not only how impacts may be distributed in time, but also how anticipated changes in towboat numbers relate to possible time-related external changes. For example, a changing climate may alter stream flow, annually and seasonally. Clearly, changes in water availability will affect both ecology and navigation. Other examples are land-use and population changes. Land-use changes can affect water quality and management requirements for ecosystem protection in the Upper Mississippi River basin. Changes in population may alter use intensity and traffic patterns. A comprehensive water resources planning analysis would consider these and other related issues in an environmental impact assessment.

Several recommendations aimed toward improving environmental analysis per se are provided in this section and discuss more thorough inclusion of those analyses in navigation project planning. The main analyses the committee focused on were a systemwide analysis of the navigation system's environmental effects, analysis of the cumulative effects of the navigation system on river ecology, and site-specific analyses.

Systemwide Analysis of Environmental Effects

There has been a long history of ecological research and analysis for the Upper Mississippi River basin, and the Environmental Management Program has established the Long Term Resource Monitoring Program (LTRMP), yet gaps in knowledge of the large, complex Upper Mississippi River ecosystem still exist. Some of those data gaps were expected to be filled by the "second lock" studies, which recognized that a detailed study is needed to characterize existing ecological status and to support impact predictions. For example, there is not a thorough understanding of how fish populations are influenced by navigation, land-use changes, and population growth in the Upper Mississippi's sub-basins. With such data gaps, full assessment of the incremental effects of proposed lock extensions on the environment is not possible. If population size and dynamics of fishes cannot be accurately assessed, how can risk predictions based on the expected mortality of fish caused by towboat passage be validated? It is simply not possible to calculate a percent loss in future population if accurate estimates of total current population size are not available. Collecting the data necessary for such a comprehensive analysis is beyond the scope of the Corps' 1997 Project Study Plan, but these data are needed to support the impact assessments required of the feasibility study. In the 1986 Water Resources Development Act, Congress established and provided initial funding for the Environmental Management Program. The Environmental Management Program has been an excellent investment of resources into the investigations of the Upper Mississippi River's ecosystems and the continuing effects of navigation improvements and other human activities. Congress should provide continued funding to the EMP, with specific and continuing appropriations for ecological studies (including cumulative effects analysis), and navigation effects studies, in the UMR–IWW feasibility study.

Cumulative Effects of the Navigation System on River Ecology

Impacts to river ecology are seldom attributable to a single cause; rather, many small effects from multiple causes typically accumulate to produce a damaging impact. A cumulative effects analysis accounts for these multiple causes. In the Upper Mississippi River Basin, the navigation system, and the operation and maintenance of its locks and dams, are affecting river ecology in multiple and complex ways. There is not a thorough understanding of how current operations (e.g., intra- and inter-annual changes in navigation pool elevations) and maintenance (e.g., dredging, construction of wing dams) of the navigation system, as well as other factors such as changes in land use and water quality, are affecting river ecology. This understanding is essential to assessing how future changes in the navigation system might affect the environment.

To address this need for enhanced understanding of these cumulative effects throughout the UMR–IWW system, a good starting point would be a detailed assessment of how current operations and maintenance activities, when combined with environmental changes, are affecting the environment. As part of the feasibility study, an analysis of the cumulative effects of the existing navigation system should be conducted. This analysis should account for project maintenance requirements based on construction methods, and—considering the duration of the project elements—it should assess maintenance and operational requirements in light of possible changes in external factors such as climate, land-use changes, or changes in population. This cumulative effects analysis should account for all major factors that can create significant environmental impacts in the UMR–IWW system. Examples of these major factors in the Upper Mississippi River Basin include spills of chemicals and fertilizers shipped on the river, changes in navigation pool elevations, and physical changes to the river channel.

An historical context may be important for some navigation system planning investigations. The USGS "Status and Trends" report (USGS, 1999), is a good example of effective use of historical data. Ongoing studies of the UMR–IWW ecosystem continue to enhance the understanding of how river ecology is changing and responding to impacts over time.

Within this cumulative effects analysis, two additional sets of more focused analysis should be conducted: recent navigation system improvements and increased towboat traffic.

Recent Navigation System Improvements. In addition to the lack of understanding of systemwide, cumulative environmental effects, there have also been concerns that the impacts of recent navigation system improvements on the UMR–IWW are not fully understood and that they have not been adequately assessed. In fact, some groups (e.g., NECC, USFWS) view continuing degradation, habitat losses, and the decline of select native species in the Upper Mississippi River basin as evidence of accumulating impacts (cf. USGS, 1999). They argue that any assessment of impacts in the feasibility study, without a better understanding of both present conditions and trends, will be unfounded. The committee agrees with this view. Cumulative effects analysis on the UMR–IWW should include a systemwide assessment that also considers the continuing effects of recent navigation system improvements (e.g., siltation of backwaters due to navigation pool formation).

Increased Towboat Traffic. Improvements to navigation systems and increases in towboat traffic will generally produce site-specific and passage-specific effects. These effects are analyzed in the Corps' draft feasibility study and are considered in the Corps' draft EIS. An inadequacy of the feasibility study is the lack of analyses that consider cumulative effects of increased towboat traffic. There is a need to establish how elements of the proposed navigation system improvements—when coupled with increases in waterway traffic—might combine to produce a total environmental impact greater than the sum of its individual parts. For example, the Corps' 1997 Project Study Plan identifies multiple locations where construction activity is anticipated in the channel. A reasonable question is, "Might multiple improvements, constructed in sequence over the life of the project, have a total impact that single or isolated projects do not?"

In the feasibility study, although a range of options and scenarios has been evaluated, consideration of the cumulative effects of increased waterway traffic is inadequate. Further, cumulative effects may have sources directly related to the presence of towboats. For example, one might ask, "What will be the cumulative effect of a spill of oil, fertilizer, or toxic chemicals on critical periods in an organism's life history, given increased traffic movement, altered pool management, or the timing of the spill?" Questions such as this have not yet been addressed in the feasibility study, yet are important to a comprehensive understanding of how multiple changes in the river system might affect the environment.

Site-Specific Impact Assessment

The impact assessments performed as a part of the feasibility study draft EIS are of necessity general at this stage of project planning. Several alternatives are still being considered; any impact analysis must thus be somewhat general. Unfortunately, this generalized impact assessment approach will often be inadequate when site-specific project activity commences. For example, detailed placement of structures, decisions on construction methods, and timing of construction activities are all uncertain at this time. Each of these factors can play an important role in actual impact. It is important to regularly review and update site-specific impact assessments as the project progresses. It is also important to recognize that each impact assessment activity can contribute valuable data to baseline data resources. It is important that impact analysis activities be coordinated with research centers so that maximum utility can be obtained from any assessment investments in the UMR–IWW system.

Detailed, site-specific impact assessments of construction activities on the UMR–IWW should be an ongoing activity in the feasibility study. It is particularly important that as experience is gained from early stages of the project, the experience contributes to better impact assessments. Furthermore, the Corps should develop a mechanism that will assure coordination with research centers in the Upper Mississippi region so that assessments will not only serve to meet impact analysis requirements, but will also contribute to improved understanding of the Upper Mississippi River ecosystem. Finally, these site-specific impact assessments should have a coordinated regulatory review. This site-specific impact assessment need not be a barrier to project progress; rather, assessments should be considered an essential element of site-specific design activities that will assure that construction effects are minimized and impacts truly avoided.

Congress should supply the resources needed to support the committee's recommendations for additional environmental investigations on the UMR–IWW. The Corps should not be required conduct these analyses alone, but rather the analyses should be conducted with the cooperation of federal and state partners who have contributed to the feasibility study. The cumulative effects and other analysis should build upon the numerous, significant research efforts on Mississippi River navigation and ecology, including the Corps' draft Environmental Impact Statement.

Toward Better Integration: Interagency Coordination

Since the 1986 Water Resources Development Act, considerable progress has been made in improving ecological understanding of Upper Mississippi River ecology. The final documentation produced by the Corps of Engineers should integrate findings from the Environmental Management Program, not only in a draft environmental impact statement, but with all elements of the feasibility study (economic, engineering, and ecological). To date, the EMP documentation stands alone, as do many of the studies performed for, and by, the Corps as part of the feasibility study. But more than an assembly of these studies is needed. Integration is not simply the collection of studies into a single report; rather, integration involves a careful analysis of findings that connects and harmonizes individual elements of the feasibility study. The current challenge to the Corps is to continue what has been started and effectively integrate all project elements within the final feasibility and environmental impact analyses. Considering the schedule and progress of the feasibility study, there has been limited time devoted to achieving final integration. The Corps should take advantage of any opportunity to secure the time needed to complete an integration of feasibility study elements with EMP studies, and with environmental analyses available from Upper Mississippi states.

Toward Better Integration: Interdisciplinary Considerations

The Corps has embarked on a landmark assessment in the feasibility study for the UMR–IWW system. Detailed economic, engineering, and ecological analyses, along with modern planning and assessment approaches, have been the foundation for this project. Although progress has been made in analytical areas, there has been little time to use the findings from one set of analyses with other analyses (economic, ecological, or engineering). It is imperative that findings from each area of analysis are effectively integrated with other analyses in final Corps feasibility and impact assessments. For example, if economic modeling suggests a change in seasonal demand that will alter the number of towboat passages along a reach of river, then specific environmental impacts associated with those changes should be assessed. If shipping schedules are altered because of economic incentives, then the altered schedules should be assessed in terms of seasonal ecosystem dynamics. If engineering plans call for winter construction, seasonal impacts of that construction should be an element of integration. These are just three of many possible examples that could be cited as needed integrative analysis in the feasibility study. The Corps should aim toward a more comprehensive and integrated assessment of the effects of changes in barge traffic on the UMR–IWW. This assessment should be consistent with National Economic Development (NED) accounting procedures and environmental quality accounting procedures.

Environmental Costs and Valuation of Ecosystem Services

An acceptable approach for addressing environmental costs would be to detail the full range of environmental impacts associated with each alternative, and to estimate economic costs associated with these impacts. Only after this analysis has been performed should the potential for mitigation and the costs and effectiveness of alternative mitigation strategies be considered. Explicit consideration of environmental costs prior to selecting mitigation strategies allows for trade-offs between unmitigated environmental costs, and the costs of mitigation to alleviate those costs, to be examined. This approach will assist in setting priorities for mitigation and in cost-effective use of mitigation funds.

Many traditional measures of productivity and the values of goods and services, such as Gross Domestic Product (GDP), do not account for the values of nonmarket goods and services. For example, the value of a forested bottomland sold as timber (market value) would be included in GDP calculations, but not that bottomland's value in attenuating floods flows or in filtering pollutants (nonmarket values). In response to this shortcoming of many traditional measures and methods, an abundant literature on the valuation of nonmarket values for environmental goods and services has blossomed.

Several methods for valuing these "ecosystem services" have been developed and include the travel cost method, the contingent valuation method, and hedonic valuation. These economic methods are well developed and are used by many public agencies. While development of these methods dates back decades (Ciriacy-Wantrup, 1947; Davis, 1964), the Exxon Valdez incident in the Gulf of Alaska in 1989 generated a great deal of interest and activities in nonmarket valuation of environmental goods and services, specifically in contingent valuation methods (Arrow et al., 1993; Carson et al., 1994). While economists and others continue to debate the utility of contingent valuation methods (Hausman, 1993), contingent valuation is a widely used tool for valuing environmental goods and services. These methods aim to understand how much individuals would be willing to pay for successive additional quantities of a nonmarket good. Although no method is foolproof, this value can be captured through questionnaires, surveys, and votes (see also Schelling, 1968). These methods today are used around the world (Hanemann, 1994) and are continually being refined. Of recent studies that have attracted attention, a group effort in the late 1990s that attempted to value all of the planet's ecosystem services and natural capital stood out (Costanza et al., 1997). This 1997 paper was quick to point out that, due to the study's complexity and ambition, that it was merely a first approximation of these values. In the field of water resources planning, there is a large and growing literature on valuation of the services of riverine and aquatic ecosystems (e.g., Bishop et al., 1987; Douglas and Taylor, 1998; Loomis, 1998).

These valuation methods, while essential in a complete examination of costs and benefits of project alternatives, cannot be usefully applied in the absence of well-defined descriptions of the range of environmental impacts that are associated with project alternatives. Moreover, these valuation methods will not be capable of quantifying (in dollars) all the environmental costs associated with alternatives. They are best suited for quantifying costs associated with environmental impacts that relate to human use of affected resources (such as boating and sport fishing) and impacts that can adequately be described to the public in the context of survey instruments of the type used in contingent valuation.

Recreation. A 1995 study sponsored by the Corps documents over 12 million visitor days annually, with associated annual economic impacts of $1.2 billion in recreation activities directly linked to uses of the UMR–IWW system (Carlson et al., 1995). Although they are not addressed in the feasibility study, recreation user benefits (consumer surplus to recreational users) are generated by these user-days (as are the economic impacts on businesses cited above). Direct benefits to recreational users are part of NED and should be considered in project evaluation.

Prior research suggests that river recreation is negatively affected by increased commercial barge traffic (Becker, 1981; Graman et al., 1984). In particular, recreational boaters respond to increased traffic by foregoing recreational boating and by relocating their activities to sites unaffected by higher traffic (with higher travel costs incurred and lower consumer surplus generated). This type of effect on recreation is consistent with studies of other forms of recreation that indicate a negative relationship between increased congestion and user benefits. Analogies can be found in the literature which documents negative impacts of increasing numbers of users at parks, fishing sites, reservoirs and hiking trails on recreation user benefits (Loomis and Walsh, 1997; pp. 104–109). The recreation economics literature clearly demonstrates that benefits per user-day fall notably as the number of encounters with other parties rises. This phenomenon should be incorporated in the analysis of project alternatives, which should explicitly model and quantify the relationships between recreation use benefits for boating and fishing and changes in commercial barge traffic levels.

River-related recreation benefits in the study area represent significant economic benefits provided by the Upper Mississippi River and are likely to be affected by many of the navigation project alternatives. The Corps should commence with a detailed analysis of impacts on recreation use benefits, going beyond the consideration of only boater safety and lock delays.

6

Summary and Recommendations

While the feasibility study represents some important advances for Corps of Engineers water resources project planning studies, many elements of the study have been framed narrowly or incompletely. Three key areas in which the draft study should be improved are 1) data and assumptions in the spatial equilibrium modeling, 2) a careful assessment of the prospects of non-structural alternatives for decreasing waterway congestion, and 3) better integration of economics and engineering considerations with environmental and social factors.

The Corps' new approach to estimating navigation benefits is a significant advance. The spatial equilibrium model is a major improvement over past models. Unfortunately, the application of the theoretical models is unsuccessful. The data and assumptions used as input to ESSENCE require considerable improvements before ESSENCE can provide reliable input into the feasibility study. The ESSENCE model simplifies the spatial equilibrium model to a point where the basic concepts are lost; the data on grain and freight shipping quantities, origins and destinations, and prices are inadequate. Values used for shipment costs and agricultural yields must reflect real variations across space.

There are also concerns about the Corps' focus on lock extensions with little consideration of nonstructural alternatives. The full range of nonstructural alternatives should be evaluated before lock extensions are considered. A comprehensive assessment of the benefits and costs of these nonstructural options for improving waterway traffic management should be conducted. Congestion management could improve waterway traffic management almost immediately, while reducing congestion by extending locks on the UMR–IWW would take a decade or more. Furthermore, nonstructural alternatives provide an excellent opportunity for the Corps to simultaneously improve waterway traffic flow and environmental resources, applying the Corps' experience in analytical methods and water resources management. The declining state of key ecological indicators on the UMR–IWW points to the need for the Corps to consider environmentally sustainable options for improving waterway traffic management.

The draft feasibility study framed the issue of waterway traffic management and possible lock extensions primarily in terms of commodity shipments and infrastructure investments. Environmental issues were treated, but mainly in terms of side effects that required mitigation. But commercial cargo represents only one of the important interests and public values on the

UMR–IWW. The draft feasibility study must better integrate environmental and social considerations into the ultimate decision of whether the locks are to be extended. Within the environmental assessments, an improved assessment study of the system-wide and cumulative effects of the existing navigation system on river ecology is needed. The Corps' progress in its environmental investigations is clearly hampered by the lack of data that such a study would provide, and new studies may be needed to meet these information needs. The federal Environmental Management Program should conduct both ecological studies and navigation effects studies; this would provide better coordination to these efforts and assure more useful input into the feasibility study.

In addition to these three broad areas for improvement, the Corps should conduct the feasibility study under an adaptive management paradigm, the study should be better coordinated with of other government agencies responsible for assessing and managing aspects of the Upper Mississippi River–Illinois Waterway system, and it should be conducted with periodical review from an independent, interdisciplinary panel of experts.

This report's recommendations should be viewed from two different perspectives: those that can be implemented immediately, and those that will require more time and a more significant commitment of resources and interagency cooperation. While some of the report's recommendations may entail sustained efforts over many years (e.g., navigation's system-wide, and cumulative, effects on the river ecosystem), others could be implemented quickly (e.g., nonstructural measures for improved waterway traffic management).

The Corps is to be commended for the advances in the draft studies. This committee offers its suggestions for improvement with full realization of the analytical complexities that the studies entail, plus an appreciation that some UMR–IWW stakeholder groups may care more about the study's recommendations than the soundness of its methods. This report's findings and recommendations are offered constructively and with appreciation for the Corps' strong efforts and significant achievements in their study of tremendous national interest and importance on the Upper Mississippi River–Illinois Waterway.

References

Anfinson, J. 1993. Commerce and conservation on the Upper Mississippi River. The Annals of Iowa 52 (4):385–417.

Anfinson, J. 1996. Henry Bosse's Views of the Upper Mississippi River. St. Paul, Minn.: U.S. Army Corps of Engineers.

Bartell, S. M, K. R. Campbell, E. M. Miller, S. K. Nair, E. P. H. Best, and D. J. Schaeffer. 2000. Upper Mississippi River–Illinois Waterway System Navigation Study: Ecological Models and Approach to Ecological Risk Assessment. Washington, D.C.: U.S. Army Corps of Engineers.

Becker, R. H. 1981. Displacement of recreational users. Journal of Environmental Management 13:259–267.

Beesley, M. E., and M. A. Kemp. 1987. Urban economics. Pp. 1023–1052 in Vol. 2 of Handbook of Urban and Regional Economics. E. S. Mills, ed. Holland: Amsterdam.

Belrose, F. C., F. L. Paveglio, Jr., and D. W. Steffeck. 1979. Waterfowl populations and the changing environment of the Illinois River valley. Illinois Natural History Survey Bulletin 32:1–54.

Carlson, B., D. Propst, D. Stynes, and R. Jackson. 1995. Economic Impact of Recreation on the Upper Mississippi River. Technical Report EL-95-16. Washington, D.C.: U.S. Army Corps of Engineers.

Carroll, J. L., and S. Rao. 1978. Economics of public investment in inland navigation: Unanswered questions. Transportation Journal 17:27-54.

Dai, M. 1993. Delay Estimation on Congested Waterways, IWR Report 93-r-8. Fort Belvoir, Va.: U.S. Army Corps of Engineers Institute for Water Resources.

de Dios Ortuzar, J., and L. G. Willumsen. 1994. Pp. 24-26 in Modelling Transport, Second Edition. New York: John Wiley and Sons.

Galloway, G. E. 2000. Three ages of river management along the Mississippi: Engineering and hydrological aspects. In New Approaches to River Management, A. J. M. Smits, P. H. Nienhuis, and R. S. E. W. Leuven, eds. Leiden, The Netherlands: Backhuys Publishers.

Graman, J. H., L. Mc Avoy, J. Abner, and R. Burdge. 1984. Relationship Between Commercial and Recreational Use of Navigation Locks on the Upper Mississippi River. Water Resources Bulletin 20(4):577-582.

Hardin, G. 1968. Tragedy of the Commons. Science 162:1243–1248.

Hart, C. A. 1895. On the entomology of the Illinois River and adjacent waters. Illinois Laboratory of Natural History Bulletin 4:149–273.

Interagency Floodplain Management Review Committee (IFMRC). 1994. Sharing the Challenge: Floodplain Management into the 21st Century. Washington, D.C.: U.S. Government Printing Office.

Jack Faucett Associates (JFA). 1997. Waterway Traffic Forecasts of the Upper Mississippi River Basin. Submitted to the Institute for Water Resources, U.S. Army Corps of Engineers. Bethesda, Md.: Jack Faucett Associates.

Jack Faucett Associates (JFA). 2000. Review of Historic and Projected Grain Traffic on the Upper Mississippi River and Illinois Waterway: An Addendum. Submitted to the U.S. Army Corps of Engineers. Bethesda, Md.: Jack Faucett Associates.

Loomis, J., and R. Walsh. 1997. Recreation Economic Decisions. State College, Penn.: Venture Publishing.

Manheim, M. L. 1979. Fundamentals of Transportation Systems Analysis. Vol. 1: Basic Concepts. Cambridge, Mass: MIT Press.

Merchants Exchange of St. Louis. 1979. Southbound Barge Freight Call Session Basis Trading Benchmarks. St. Louis, Mo.: Merchants Exchange of St. Louis.

Meyer, J. R., and M. R. Straszheim. 1971. Pricing and project evaluation. Pp. 99-109 in Vol. 1 of Techniques of Transportation Planning. J. R. Meyer, ed. Washington, D.C.: The Brookings Institution.

National Research Council. 1999a. Downstream: Adaptive Management of Glen Canyon Dam and the Colorado River Ecosystem. Washington, D.C.: National Academy Press.

National Research Council. 1999b. New Directions in Water Resources Planning for the U.S. Army Corps of Engineers. Washington, D.C.: National Academy Press.

National Research Council. 2000. Risk and Uncertainty in Flood Damage Reduction Studies. Washington, D.C.: National Academy Press.

Nelson, J. C., R. E. Sparks, L. DeHaan, and L. Robison. 1998. Presettlement and contemporary vegetation patterns along two navigation reaches of the Upper Mississippi River. Pp. 51–60 in Perspectives on the Land-Use History of North America: A Context for Understanding Our Changing Environment, T. D. Sisk, ed. U.S. Geological Survey Report USGS/BRD/BSR-1998-0003. Reston, Va.,: USGS. Also http://www.nbs.gov/luhna/-emtc/index.html.

Nyberg, J. B. 1998. Statistics and the practice of adaptive management. In Statistical Methods for Adaptive Management Studies. V. Sit and B. Taylor, eds. Land Management Handbook 42. Victoria, British Columbia: Ministry of Forests.

Reebie Associates. 1999. Rail Cost Analysis Model. Reebie Associates Transportation Management Consultants, Stamford, Connecticut.

Scarpino, P. 1985. Great River: An Environmental History of the Upper Mississippi, 1890–1950. Columbia, Mo.: University of Missouri Press.

Small, K. A., and C. Winston. 1999. The demand for transportation: Models and applications. Pp. 11-55 in Essays in Transportation Economics and Policy: A Handbook in Honor of John R. Meyer. Washington, D.C.: The Brookings Institution.

Sparks, R. E. 1977. Environmental inventory and assessment of navigation Pools 24, 25, and 26, Upper Mississippi and Lower Illinois rivers. An electro-fishing survey of the Illinois River. Illinois Water Resources Center Special Report No. 5. UILU-WRC-77-0005: pp. 1–122.

Thomas, E. N., and J. L.Schofer. 1970. Strategies for the Evaluation of Alternative Transportation Plans. NCHRP Report 96. Washington, D.C.: Transportation Research Board.

U.S. Army Corps of Engineers (USACE). 1991. Plan of Study: Navigation Effects of the Second Lock—Melvin Price Lock and Dam. St. Louis, Mo.: U.S. Army Corps of Engineers.

U.S. Army Corps of Engineers (USACE). 1997a. Report to Congress: An Evaluation of the Upper Mississippi River System Environmental Management Program. Rock Island, Ill.: U.S. Army Corps of Engineers.

U.S. Army Corps of Engineers (USACE). 1997b. Upper Mississippi River–Illinois Waterway Navigation Study. Project Study Plan. Rock Island, Ill.: U.S. Army Corps of Engineers.

U.S. Army Corps of Engineers (USACE). 1998. A Spatial Price Equilibrium Based Navigation System NED Model for the UMR–IW Navigation System Feasibility Study. St. Louis, Mo.: U.S. Army Corps of Engineers.

U.S. Army Corps of Engineers (USACE). 1998. An Overview of Benefit Estimation Procedures. Huntington, West Va.: U.S. Army Corps of Engineers.

U.S. Army Corps of Engineers. 2000a. Planning Guidance. ER 1105-2-100. Washington, D.C.: U.S. Army Corps of Engineers.

U.S. Army Corps of Engineers. 2000b. Upper Mississippi River–Illinois Waterway System Navigation Study Feasibility Report (draft). Rock Island, Ill.: U.S. Army Corps of Engineers.

U.S. Department of Agriculture. 2000. Agricultural Baseline Projections to 2009. Prepared by the Interagency Agricultural Projections Committee. Staff Report No. WAOB-2000-1. Washington, D.C.:U.S. Department of Agriculture.

U.S. Geological Survey. 1999. Ecological Status and Trends of the Upper Mississippi River System 1998: A Report of the Long Term Resource Monitoring Program. La Crosse, Wis.: U.S. Geological Survey.

U.S. Water Resources Council. 1973. Water and Related Land Resources: Establishment of Principles and Standards for Planning. Federal Register 38:24784, 248222–248223.

U.S. Water Resources Council. 1983. Economic and Environmental Principles and Guidelines for Water and Related Land Resources Implementation Studies. Washington, D.C.: U.S. Government Printing Office.

Upper Mississippi River Basin Commission (UMRBC). 1982. Comprehensive Master Plan for the Management of the Upper Mississippi River System. Minneapolis, Minn.: Upper Mississippi River Basin Commission.

Upper Mississippi River Conservation Committee (UMRCC). 2000. A River that Works and a Working River. Rock Island, Ill.: Upper Mississippi River Conservation Committee.

Volkman, J. M., and McConnaha. 1993. Through a glass, darkly: Columbia River salmon, the Endangered Species Act and adaptive management. Environmental Law 23:1249–1272.

Water Resources Development Act of 1986. U.S. Public Law 99-662, Nov. 17, 1986, 100 Stat. 4082. 99th Congress, 2nd Session.

Appendixes

Appendix A

Participating Agencies in the Corps' 1991 Plan of Study

Federal agencies:

U.S. Army Corps of Engineers (Rock Island, St. Louis, and St. Paul district offices, and the Waterways Experiment Station, Vicksburg, Mississippi)
U.S. Environmental Protection Agency (Kansas City regional office)
U.S. Fish and Wildlife Service (Environmental Management Technical Center and Rock Island field office)

State agencies:

Illinois Department of Conservation
Illinois State Water Survey
Iowa Department of Natural Resources
Minnesota Department of Natural Resources
Missouri Department of Natural Resources
Wisconsin Department of Natural Resources

Appendix B

U.S. Fish and Wildlife Service Executive Summary for the Corps of Engineers System Navigation Study

August 31, 2000

Colonel William J. Bayles
District Engineer
U.S. Army Engineer District, Rock Island
Clock Tower Building, P.O. Box 2004
Rock Island, Illinois 61204-2004

Dear Colonel Bayles:

The U.S. Fish and Wildlife Service (Service), Rock Island Field Office, is preparing a Draft Fish and Wildlife Coordination Act (FWCA) Report for the Corps of Engineers (Corps) Upper Mississippi and Illinois Rivers System Navigation Study (System Study). Mr. Ken Barr, Chief of your Environmental Analysis Branch, has requested that we provide you with the preliminary conclusions and recommendations of the forthcoming FWCA report. This letter responds to that request. However, it does not necessarily represent the Secretary of Interior's final position on the proposed project or that of the five state natural resource agencies, which include Illinois, Iowa, Missouri, Wisconsin, and Minnesota (state agencies). Our comments apply to the preliminary array of alternatives provided to us, since the Corps has yet to identify a recommended plan. Furthermore, our conclusions and recommendations are preliminary and subject to change because they are based on project information available as of May 2000.

PROPOSED PROJECT

Some of the current array of project alternatives could result in improvements at up to 11 locks and dams on the Mississippi River and two on the Illinois River. These improvements include small-scale improvements, such as mooring cells and extended guidewalls, and large-scale improvements such as 1200-foot lock chambers. As of this date, no alternative has been identified as preferred by the Corps. All six of the preliminary alternatives are predicted to reduce lockage time and allow increased navigation traffic on the Upper Mississippi River System (UMRS). The predicted incremental increase in tow traffic that would be caused by the proposed improvements is the sole focus of the Corps' environmental analysis. The "No Action" alternative represents the existing and future traffic levels that will occur without any additional navigation improvements.

ENVIRONMENTAL SETTING

The proposed project will affect one of our country's most significant ecosystems. The UMRS is a complex mosaic of bottomland forests, wetlands, and aquatic habitats which are home to over 150 species of fish and 44 species of Unionid mussels. Over 40 percent of North America's migratory birds use the Mississippi Flyway. There are 265,000 acres of the National Wildlife Refuge system along the Upper Mississippi River mainstem, and another 10,000 acres along the Illinois River. The five state agencies manage over 140,000 acres on the UMR and the Illinois DNR manages approximately 60,000 acres on the Illinois River.

ENDANGERED SPECIES ACT COMPLIANCE

The Service and the Corps recently completed formal consultation under Section 7 of the Endangered Species Act which assessed the effects of the existing Nine-foot Channel Navigation Project upon seven federally listed endangered and threatened species. These are the pallid sturgeon, Higgins' eye pearly mussel, winged maple leaf mussel, Indiana bat, bald eagle, decurrent false aster, and least tern. The Service determined that the continued existence of two species, the Higgins' eye pearly mussel and pallid sturgeon, would be in jeopardy if reasonable and prudent alternatives were not implemented by the Corps. The Service and the Corps are still consulting with respect to the effects of the proposed system-wide navigation improvements described in the System Study. The recently completed consultation for the existing Nine-foot Channel Navigation Project will serve as the baseline condition for the proposed improvements consultation.

SUMMARY OF NAVIGATION STUDY ISSUES AND CONCLUSIONS

The current analysis is deficient.

Identifying and quantifying the environmental effects of commercial navigation traffic on UMRS natural resources has been a controversial issue for more than 25 years. During this time several efforts have investigated one specific navigation effect or another, but a comprehensive evaluation of all navigation-related effects has yet to be accomplished. Such an evaluation would examine the direct and indirect effects of passing towboats on natural resources as well as the effects of operation and maintainance (O & M) of the Nine-foot Channel Navigation Project (e.g., dredging, water level regulation, channel regulatory structures, impoundment). Since the current investigation has been presented as a "system-wide" study designed to identify needed improvements throughout the entire UMRS, the Service and state agencies believe that the system-wide cumulative effects of all Nine-foot Channel Navigation Project related activities should be examined as well.

There are five major deficiencies with the Corps of Engineers environmental impact analysis. Two of these deficiencies concern the Corps' incremental traffic effects analysis (Items 1 and 2 below), and three deficiencies involve related navigation effects issues (Items 3 to 5 below). The deficiencies are as follows:

1. The assessment of incremental traffic effects associated with the proposed project is inadequate; site-specific assessments are not complete.

<u>Incremental Traffic Effects Analysis</u>

The impact analysis has resulted in the development of state-of-the-art navigation traffic effects models. Despite the development of these tools, critical biological and physical information needed to run these models is lacking. Limited knowledge of UMRS species life histories has severely hampered scientists' ability to quantify navigation effects with any degree of certainty. Because of limitations in biological and physical data, fish and wildlife impacts attributable to incremental traffic increases cannot be adequately predicted or quantified at this time.

Insufficient time was allotted in the feasibility study to collect the necessary physical, chemical, and biological data in order to quantify with any certainty impacts to fish, mussels, aquatic plants, bottomland hardwoods, and other resources. Our conclusions and recommendations regarding detailed effects due to the incremental increase in traffic are included as Table 1 enclosed with this letter.

<u>Site-specific Impacts of Proposed Construction</u>

There is insufficient lock and dam design information available at this time to complete a full evaluation of impacts to natural resources resulting from site-specific construction activities (including 1200-foot locks, extended guidewalls, wing dikes, and mooring facilities). Significant resources which may be affected by construction include fisheries, mussel beds, bottomland forests, main channel border, non-forested wetlands, Federal and state-listed endangered species, and recreational use areas.

Although preliminary assessments have been conducted, impacts to resources will be investigated in greater detail when separate environmental assessments are completed for each proposed construction site. Additional evaluations which will be necessary include Habitat Evaluation Procedures (HEP), mussel surveys, hydraulic modeling, and screening for endangered and threatened species. Potential impacts associated with the release of contaminants due to construction have not yet been evaluated and will require further investigation.

2. An assessment of baseline traffic effects which will result from the "No Action" alternative is needed to understand the effects of additional traffic.

In this System Study, the Corps investigation of systemwide navigation impacts focused only on fish and wildlife effects associated with traffic increases attributable to the proposed alternatives. This approach continues the Corps' practice of addressing the various environmental effects of navigation improvement actions in isolation from one another. The Service does not support this approach. It does not appear to comply with the National Environmental Policy Act (NEPA) because it fails to consider the cumulative effects of all navagation project related actions.

The impact analysis is so focused on incremental traffic effects, that the fish and wildlife impacts associated with the baseline traffic condition ("No Action" alternative) are not presented. Without the ability to compare the impacts of the baseline traffic condition against the impacts of the proposed navigation improvement alternatives, we can only speculate about the true environmental significance of the Corps' proposed navigation improvements. Consequently, it is even more difficult to arrive at a mitigation plan which can be linked to incremental traffic effects in any credible manner. According to Corps' policy, as we understand it, baseline traffic effects are not subject to mitigation within the definition of this study. However, the fact remains that baseline traffic continues to degrade fish and wildlife resources. In spite of ongoing habitat improvement programs (e.g., Environmental Management Program), the long-term ecological integrity of the river's fish and wildlife resources still appears to be in decline. This conclusion is supported by the 1998 Ecological Status and Trends Report prepared by the Long Term Resource Monitoring Program (LTRMP). We are particularly concerned about impacts to National Wildlife Refuge system lands and state-managed areas.

3. An assessment of impacts from operation and maintenance activities is needed.

The Navigation Environmental Coordinating Committee (NECC) has repeatedly recommended that the System Study include an assessment of the operation and maintenance (O&M) impacts associated with the existing navigation project. Despite repeated requests up through the Department level, the issue of assessing O&M impacts within the System Study is still unresolved. As we have indicated on numerous occasions, unless the Draft Environmental Impact Statement (DEIS) contains a thorough analysis of these effects, it will be a technically deficient document.

The Corps has determined that assessment and mitigation of Nine-foot Channel Project O&M effects are not applicable to this System Study. We disagree. Based on the LTRMP 1998 Status and Trends Report, we conclude that the long-term effects from O&M activities probably pose a more serious threat to UMR fish and wildlife resources than future incremental traffic effects identified thus far. Further evidence that ongoing navigation project O&M activities are causing habitat degradation is found in the Service's recently completed Biological Opinion prepared for the UMR Nine-foot Channel Navigation Project endangered species consultation. That consultation concluded that continued O&M activities are causing, and will continue to cause, such a decline in Higgins' eye pearly mussel and pallid sturgeon populations that, unless remedial actions are immediately implemented, their future existence on the UMRS is jeopardized.

Some O&M effects have never been addressed in any Corps of Engineers National Environmental Policy Act document. O&M effects that the Corps should investigate further include: (1) facilitated spread of exotic species; (2) habitat degradation from water level regulation; (3) loss of habitat diversity from construction of channel regulatory structures; (4) habitat degradation from ongoing sedimentation due to impoundment; and (5) hindrance of fish passage through navigation dams. These O&M impacts are addressed in more detail in our forthcoming draft FWCA report and supporting appendices. We are likely to conclude that the NEPA analysis found in the draft Environmental Impact Statement for the Systemic Navigation Study is inadequate if these effects are not included in the analysis.

4. An assessment of traffic impacts and mitigation for the Second Lock was never completed.

Traffic-related impacts resulting from the Second Lock at Melvin Price Lock and Dam have not been quantified. The Corps is proposing additional navigation improvements before the effects of Second Lock improvements are known. (NOTE: The Service has been waiting for such information since the late 1980's in order to prepare a final FWCA report for the Second Lock Project). Since initiation of the System Study, the Corps has declined to assess and determine appropriate mitigation for Second Lock traffic impacts. Traffic from the Second Lock generates the same types of impacts as the additional traffic from the proposed project. The traffic from the Second Lock is, in fact, considered part of the baseline traffic condition for the System Study. In order to quantify the impact of the incremental traffic increases associated with the proposed project, impacts attributable to the Second Lock increment of traffic must be identified and mitigated for as part of the current System Study.

In the 1980's, there was considerable disagreement over the navigation effects of the Second Lock. The Service, as well as the U.S. Environmental Protection Agency, state agencies, and several environmental organizations were promised that if the Corps was allowed to construct the Second Lock, the Corps would undertake a full investigation of traffic effects. Such an investigation, described in what became known as the Plan of Study (POS), was designed with the full participation of all agencies and finalized in 1992. In addition, the Corps committed to implementing a program to avoid and minimize Second Lock traffic effects to fish and wildlife throughout the UMRS. The Corps has failed to implement both of these commitments which were contained in the Second Lock Record of Decision. The POS was incorporated into the current System Study impact analysis, but was so reduced in scope that the uncertainty of the results leaves us far short of any comfortable assurance level that fish and wildlife resources will be protected or compensated. Only the St. Louis District has implemented an avoid and minimize program.

5. A Comprehensive Mitigation Plan to address all Nine-foot Channel Navigation Project effects is lacking.

We endorse a mitigation approach that offers flexibility, yet commits the Corps to an authorized level of funding regardless of future traffic levels. Our recommendations for adaptive mitigation have been previously described in the draft FWCA report for the Second Lock at Melvin Price Lock and Dam. An adaptive approach would allow us to select the most appropriate measures for impact locations along the UMRS. For an adaptive approach to be successful, the Service and state agencies must be closely involved in the design and selection of measures. There must also be a mutually agreeable coordination procedure established between the Corps, the Service, and the state agencies to implement an adaptive mitigation plan. This could be accomplished by a committee co-chaired by the Corps and Service, with the assistance of the existing St. Paul District's River Resource Forum, the Rock Island District's River Resources Coordinating Team, and a similar group in the St Louis District. Most importantly, there must be adequate assurance that an adaptive approach will actually be implemented.

We recognize the Corps' immediate need to specify a mitigation cost for calculating overall project costs and benefits. The Service believes the mitigation costs cannot yet be determined because there are too many data gaps. The Service and state agencies have been asked to accept a fixed mitigation cost in spite of the fact that there is no strong biological linkage between the impacts and the proposed measures. The assignment and scale of these measures to specific incremental traffic effects was prepared with little input from the Service and state agencies. The Service and the state agencies have had insufficient involvement in determining which measures should be applied, and to what degree, for a given impact. For example, in order to mitigate systemic fish impacts, the Corps proposed a $40 million fish passage measure at L/D 19 without coordinating with the natural resource agencies. In spite of our collective non-support for this measure, it is still part of the mitigation plan.

Most of the proposed measures will generate habitat improvements, but no methodology was used to demonstrate that identified costs of measures will result in net habitat gains equal to the corresponding level of loss. The net gain in habitat from these measures is speculation at this point. For example, how many larval fish would be generated by a fish passage structure at L/D 19? Normally, an incremental analysis impact planning tool (e.g HEP, WHAG) is employed to determine the type and appropriate amount of mitigation needed to offset each impact. Such traditional tools are not readily applicable to incremental traffic effects.

This situation would indicate that an innovative approach, requiring intense coordination among the Corps, the Service and state agencies, is needed to prepare a biologically credible mitigation plan. For this reason, the Service has recommended that an adaptive mitigation approach be used. The adaptive approach would involve implementing mitigation measures on a small scale, monitoring their habitat benefits, and making refinements before it is implemented systemwide. The Corps has indicated a willingness to consider an adaptive approach to mitigation as stated in their December 21, 1999 draft document Adaptive Mitigation Implementation Strategy presented to the NECC. A critical deficiency in this draft document is the lack of a formal process for implementing the adaptive approach. Most importantly, the specific roles and responsibilities of the Service and state agencies must be described.

Thus, the proposed mitigation plan is unacceptable because: (1) incremental impacts have yet to be quantified with any certainty; (2) a strong biological link between the quantity of impacts identified with the quantity of habitat improvements proposed has not been established; (3) there is no documentation of any incremental analysis used to illustrate how the proposed mitigation measures were selected; (4) there is no implementation scheme with respect to how mitigation would be accomplished; and (5) the mitigation plan does not include remedial measures for Second Lock impacts, ongoing O&M effects, or baseline traffic.

The incremental traffic mitigation package is presented independent of all other navigation traffic effects. By mitigating for the Nine-foot Channel Navigation Project as a whole, we can avoid duplication and fragmentation. It would permit us to remediate impacts appropriately for fish and wildlife that depend on the system as a whole. The incremental traffic mitigation package should be integrated with mitigation for other navigation effects (e.g., O&M effects, baseline traffic, Second Lock) into one system-wide comprehensive plan.

Site-specific mitigation plans have not yet been developed. Estimated habitat replacement costs are currently $18.6 million. We are currently unable to determine whether or not this will be adequate since site-specific designs are yet to be finalized.

RECOMMENDATIONS

1. Incremental and Site-Specific Traffic Effects

A. Incremental Traffic Effects

More credible and accurate predictions of incremental traffic effects must be prepared before an acceptable mitigation program can be developed. In order to achieve this, the following data gaps must be addressed.

1. Collect system-wide bathymetric and sediment data to improve hydraulic model accuracy.

2. Reinitiate the main channel fish entrainment investigation responsive to the recommendations previously given by the Service. Impact models should then be run again to quantify navigation effects on main channel fish.

3. The plant growth models used to determine sediment effects on plants need to be verified against other native submergent plant species growth characteristics. In addition, locations of other submergent vegetation patches downstream of Pool 13 and in the Illinois River must be mapped and evaluated.

4. Preliminary findings that Unionid mussels will not be significantly affected must be verified through further investigation to assess reproductive effects and physiological effects with respect to glycogen content and cellulose activity. In addition, effects of increased traffic related to zebra mussel distribution should be assessed.

5. Impacts associated with the increased need for fleeting and commercial river terminals need to be quantified. The Corps should undertake a comprehensive planning effort to identify suitable and unsuitable navigation development locations.

6. Impacts to National Wildlife Refuge lands and state lands should be identified.

B. Site-Specific Effects

1. The HEP analyses previously conducted for Locks 20-25, Peoria, and LaGrange should be revised and re-evaluated consistent with proposed design plans. Additional HEP analyses should be initiated and completed for all other proposed construction sites.

2. Potential impacts to fishery resources should be evaluated in greater detail using both HEP and hydraulic modeling at all proposed construction sites.

3. Results of previous mussel surveys should be reviewed to determine where additional surveys are needed. Measures to avoid and minimize harm to mussels must be incorporated into the design plans.

4. Current lists of state endangered and threatened species potentially affected by construction should be expanded to include all known records of occurrence at the county level. Species identified in the county lists should be presumed to potentially occur in the immediate vicinities of the corresponding locks and dams unless they can be ruled out based on habitat. Field sampling to determine the presence or absence of species should be conducted where appropriate.

5. The potential for release of contaminants resulting from construction activities should be assessed for each proposed construction site. More detailed investigations should then be conducted at sites with high potential for release.

6. Impacts to recreational opportunities should be analyzed in connection with loss of aquatic and terrestrial habitats, loss of boat ramps, loss of access to areas currently fished (including banks and areas behind guidewalls), and loss of access to wildlife viewing areas.

2. Baseline ("No Action") Traffic Effects

A. The traffic impacts associated with the project baseline condition must be quantified and presented along with the incremental traffic effects analysis.

B. A mitigation plan to offset baseline traffic effects should be included in a system-wide comprehensive plan to mitigate all Nine-foot Channel Navigation Project related impacts.

3. Cumulative Effects of O&M Activities

A. The Corps of Engineers should complete an updated assessment of system-wide O&M impacts (e.g., sedimentation, spread of exotic species, alteration of the natural hydrograph from dam regulation, channelization, shoreline erosion, and fish passage) associated with the existing Nine-foot Channel Navigation Project. This should be completed as part of the current System Study.

B. Coupled with this analysis should be the development of a system-wide program to avoid and minimize these impacts. This effort should include, but not necessarily be limited to, the following:

1. Alternative measures to improve fish passage at navigation locks and dams should be evaluated and implemented if appropriate.

2. The Corps' channel regulatory works program should investigate and implement mitigation techniques that can both maintain a navigation channel and restore aquatic habitat diversity.

3. Navigation dam water regulation procedures need to be investigated and revised to allow for fish and wildlife considerations. Alternative water level manipulation protocols need to be developed to determine optimum drawdown scenarios. This should include development of a drawdown schedule that balances fish and wildlife needs with commercial navigation and public use.

4. The Corps should immediately seek authority to manage its navigation project lands and waters for the benefit of fish and wildlife as a project purpose equal to navigation. This equal authority should include the ability to implement fish and wildlife management actions deemed critical to the long-term preservation of fish and wildlife and their habitats.

5. The Corps of Engineers should investigate the feasibility of eliminating the waterway connection between Lake Michigan and the Illinois Waterway. Alternatives to the existing connection should be investigated in order to sever the link that has become a pathway for the movement of exotic species between the Great Lakes and the Mississippi River Basin.

4. Second Lock Traffic Impacts

A. As part of this feasibility study, the Corps should quantify the increment of traffic attributable to the Second Lock at Melvin Price Lock and Dam and, in consultation with the Service and state agencies, include appropriate mitigation (if any) in the current System Study.

B. An avoid and minimize program, as described in the Second Lock ROD should be implemented system-wide in conjunction with a comprehensive mitigation plan.

5. Mitigation

A. Continue funding support to the Environmental Management Program (EMP). Without the EMP, the current level of knowledge concerning navigation impacts gained thus far would not have been possible. The EMP's Habitat Rehabilitation and Enhancement Program should be considered as a means to evaluate the effectiveness of different types of measures that could mitigate Nine-foot Channel Navigation Project effects. Monitoring and evaluation of mitigation measures should be integrated into the LTRMP so that, habitat values can be maximized and dollar costs minimized.

B. Implement an adaptive mitigation approach that comprehensively mitigates for all incremental traffic effects as well as the Second Lock, baseline traffic, and O&M effects. Because of the extreme uncertainty associated with future impacts, a strategy that allows flexibility in implementing mitigation actions is necessary. The adaptive approach should include the ability to monitor mitigation actions to evaluate their effectiveness for future actions. Therefore, we recommend that an adaptive mitigation strategy be developed that includes the following:

1. Since the majority of navigation effects are anticipated to occur during the latter half of the fifty-year project life, funding for a comprehensive mitigation plan should be provided by a trust fund that allows for funding over the fifty- year life of the project. This is necessary to assure the availability of funding after the completion of the proposed improvements.

2. An independent panel of scientists should be convened to give assistance in preparing a General Plan to: (a) improve the quantification of incremental, Second Lock, O&M, and baseline traffic effects; and (b) develop a strategy to implement an adaptive mitigation approach.

3. A Navigation Mitigation Steering Committee consisting of the Corps, Service, and state agencies should be established immediately. The Committee should be co-chaired by the Rock Island District and the Rock Island Ecological Services Field Office. The Committee will provide direction to adaptive mitigation

implementation but should utilize the existing fish and wildlife work groups (e.g., River Resources Forum in the St. Paul District, the River Resources Coordinating Team in the Rock Island District, and any similar group in the St. Louis District) for development and implementation of specific projects. Membership should include the state agencies and the U.S. Environmental Protection Agency. Voting members should include one each from the following: a) Corps of Engineers; b) Fish and Wildlife Service; c) Iowa Department of Natural Resources; d) Illinois Department of Natural Resources; e) Minnesota Department of Natural Resources; f) Missouri Department of Conservation; and g) Wisconsin Department of Natural Resources.

4. The Committee should monitor work being done through the EMP to insure that applicable information is being incorporated into the adaptive mitigation planning process. The Corps, in co-leadership with the Service, should complete a General Plan for implementing adaptive mitigation. The plan should use available information obtained from the EMP's habitat rehabilitation and long-term monitoring programs as well as additional information from studies needed to fill data gaps recognized in this investigation. This General Plan should be endorsed by the Committee. The adaptive mitigation concept will develop short-term (5-year) mitigation implementation plans. As one term comes to an end, the Corps, with assistance from the Service and in cooperation with the Committee, will evaluate the effectiveness of mitigation measures. This information, along with new data, will be used to develop the mitigation implementation plan for the next term. The goal is to always have full compensation of unavoidable habitat losses.

5. The General Plan will address actions which should be taken concurrent with navigation improvement construction. The General Plan should also address the mitigation funding requirements and allocations so that funding is insured throughout the life of the project and achieves the greatest mitigation value. The plan should be updated as the results of monitoring studies become available.

6. An initial mitigation implementation plan should be developed under the co-leadership of the Corps and the Service. It should be endorsed by the Committee. Primarily, the first term plan should implement all feasible measures to avoid and minimize impacts.

7. The District should complete all subsequent plans in sufficient time to be entered into the Corps' budget cycle and insure adequate funding. Compensation to mitigate adverse impacts should not be considered until all practicable measures to avoid and minimize impacts have been implemented.

We believe the above recommendations provide a flexible solution to a very difficult problem. However, a strong commitment from the Corps of Engineers will be necessary to bring this program to fruition. Without such a commitment, we can only conclude that the impacts of any increases in navigation traffic will not be adequately mitigated, and balanced use of this nationally significant ecosystem will not be achieved.

Of all the issues discussed in this letter, mitigation planning is possibly the most urgent. We need to increase our coordination efforts to: (1) develop a detailed mitigation implementation strategy and (2) develop a method to link impacts and mitigation measures. The Service recommends that "workshop meetings" of the NECC be scheduled soon to address these two issues. This recommendation is supported by the NECC representatives from all the state agencies. Please feel free to contact me or Mr. Jon Duyvejonck of my staff to continue our coordination on this or any other navigation issues.

This letter provides comments under the authority of and in accordance with provisions of the Fish and Wildlife Coordination Act (48 Stat. 401, as amended; 16 U.S.C. 661 et seq.); and the Endangered Species Act of 1973, as amended.

Sincerely,

Richard C. Nelson
Supervisor

Enclosure

cc: NECC

Appendix C

Tradable Lockage Permits

The job of smoothing arrivals at locks should be given to towboat operators, as they best understand the costs of delay and the technical limitations inherent in the operation of their equipment. The most direct way to achieve the smoothing of arrivals is to give towboat operators clear property rights to lockage times at different locks. This would be done by investigating the historical pattern of lock usage and, before the beginning of the navigation season, allocating to each operator the same percentage of 5-minute blocks throughout each day as they used in the historical period. It is important that the 5-minute blocks be allocated evenly over the year, with each operator being given the same number of minutes in each day in August as in April or October. The 5-minute slots throughout each day would be allocated at random to each operator. The remainder of the lock minutes throughout each day would be retained by the U.S. Army Corps of Engineers. Before the beginning of the navigation season, any operator could swap an assigned 5-minute slot for one held by the Corps.

The key to demand smoothing is to encourage the swapping of slots among operators to assemble a clear sailing path through each lock. Thus, for example, an operator planning an upbound sailing would swap with others to gain clear title to a lockage time at Lock 25 that is 6 hours, sailing time from the lockage time owned at Lock 26. The Corps would sell the minutes that it retains at the beginning of the navigation season to the highest bidder. Bidding would continue p to the point when the lockage time slot expired. This would allow operators a fall-back position in case they failed to reach a lock in time to use the time slot that they owned. The Corps should develop rules for reserving slots for noncommercial lockages as well.

If an operator stayed in a lock for longer than the time that had been reserved, an arbiter would determine the penalty. In principle, an operator should be liable for all delays imposed on other users. Thus, if contiguous lockage times were delayed by an operator's failure to clear the lock on time, the operator should be liable for delays imposed on all those holding slots behind him or her as well as for delays that the others might encounter at the upstream or downstream sequence of lockages that they were scheduled to traverse. However, although the length of time a boat spends in a lock is determined primarily by the towboat operator, it is partially under the control of Corps employees. Weather also might cause a lockage to take longer than anticipated.

Under these circumstances, an arbiter would be permitted to forgive some of the fees that would normally be imposed on an operator who took longer in a lock than the time reserved.

New operators would buy lockage slots from either existing operators or from the Corps. To encourage entry, however, at the end of the navigation season, the Corps should be directed to rebate back to the new entrant an amount equal to the value of time per sailing given to existing operators.

Allowing operators clear title to specific 5-minute blocks of time throughout each day of the navigation season and then allowing them to trade slots to schedule their own lockage sequences is the simplest, cheapest, and most direct way to speed the flow of traffic on the UMR–IWW system. It has the added advantage of increasing the speed of lockages and thus increasing the system's economic capacity.

Implementing Tradable Permits—Experience from Other Programs

A tradable permit approach holds promise for inducing more efficient use of navigation infrastructure by the barging industry on the Upper Mississippi River–Illinois Waterway system. Tradable time-specific permits to use locks may reduce congestion delays and other user costs and may encourage technological innovation. In considering this approach, it is worthwhile to note that implementing tradable permit policies is often highly politicized because of the complex array of stakeholders with an interest in a tradable permit program. There are those who make a direct livelihood from the navigational use of the waterways—barge captains and tow operators; and those linked to them through capital investments, input supply, and output processing—from grain elevators and port authorities, and from communities concerned with jobs and economic vitality. There are competing advocates for differing priorities in waterway management: recreational boaters, national and local environmental advocates, and farm organizations. There are multiple arms of government—the Corps of Engineers, federal and state wildlife agencies, Congress, and the courts.

These disparate interests are faced with multiple tasks. Prior to establishing tradable lockage permits, policymakers must determine the desirable and feasible number of commercial navigation lockages per unit of time at specific locks or series of locks, and they must consider exceptions to the permitted lockage times for unusual conditions (e.g., severe weather). The nature of the permits must be specified. The permits could be in the form of a right to make an advance reservation that would avoid time in queue at specific locks. Or lockage arrival times and time allotted within the lock could be pre-assigned based on historic dates of use and use levels, with trading to adjust to current needs. To whom would the "permits" be allocated? Should a permit be mandatory to use a lock, or required only to move to the front of the queue upon arrival? What penalties apply for failure to use an assigned lockage time or for use of excess time in the lock? Should the program be revenue-neutral—that is, fees collected would be returned to the permittees in a manner that does not distort their lockage use incentives.

Some organization (which could be a public agency or a voluntary association formed by navigation interests) must assign initial lockage permit allocations, set the rules for trading, and monitor and enforce the activities that ensue. A central coordinating system for reservations and/or trading might involve round-the-clock dispatchers, communicating via radio or the Internet. Fees collected could be dedicated to such administrative and coordinating costs. Finally, when all the details are worked out and the system begins to operate, policymakers must account to stakeholders, the public, and (frequently) the courts for the resulting stream of impacts on various constituencies.

Economists' advocacy of tradable permits has been an invaluable policy contribution, and tradable permit programs provide genuine advantages over command-and-control allocation. The development of markets in SO_2 allowances, western U.S. water rights, and fishery individual transferable quotas (ITQs) provide instructive insights for development of a tradable lockage permit program.[1] Development of tradable permits for these three resources is briefly described in the appendix to this section.

Establishing tradable permits entails a series of policy design challenges when permits are defined and allocated and rules for trading are developed. This section outlines those challenges, based on experience with other tradable permit programs such as the three described in the appendix, and their implications for tradable lockage permits for navigation waterways.

New Entrants

The political complexity entailed in crafting a trading program often causes delays from the time tradable permits first are publicly debated to the date that trading rules are promulgated and permits are allocated. In the interim, all those with even a minor interest in the resource seeks to shore up their claim. Fish catch, pumping levels, and emissions soar in anticipation of rights being granted based on a history of use, anticipating the windfall gains that will go to those granted tradable rights. Alaskan fisheries amply demonstrated a rush on the resource in the face of pending quota policies.

The issue of how to incorporate new navigational interests into an established system of lockage permits will need to be addressed. New entrants (and increased use by an existing permittee) could be accommodated by allocating them lockage permits at zero cost. This approach would lessen incentives for trading, as compared to requiring new users to buy a permit. However, if permits issued after the program is established are for the less heavily used (and less desirable) dates of the navigation season, then trading likely would occur between those with permits for more desirable dates and those with less desirable dates.

Quantifying Tradable Rights

The "rights" being traded in fishery, air emissions, and water markets (see appendix) had begun to evolve before tradable permits were proposed. Resource users had customary patterns of use, implicitly sanctioned by public policy. When pressures on water supply, fish stocks, and air quality became sufficiently urgent, there was momentum to quantify rights and develop trading mechanisms.

---

[1] This report only briefly describes each case. There is richly detailed literature on each of these three markets to which readers are referred for in-depth description and assessment.

At this point in time, there does not appear to be a mandate to cap navigation use at any specific level. Presumably, lockage permits would be allocated to all current navigation users of the waterways. The question arises regarding what quantity each applicant would be awarded and who would be eligible for an allocation. One can envision a situation similar to the Alaska fisheries in which permits are awarded far in excess of use levels in recent years, in order to minimize political controversy. If tradable lockage permits are awarded in excess of actual use levels, incentives to trade will be weakened, though trading would still occur to obtain more desirable lockage time slots and to accommodate unforeseen delays that prevent the use of an advanced reservation.

As the UMR–IWW system is currently managed, navigation users have the "right" to arrive at locks without any reservation, permit, or coordinating activity of any type required. At some locks, congestion costs under the current system are considerable. It is the opportunity to reduce these costs that would motivate support for, and participation in, some form of reservation or permit system. The intensity of use of a reservation or permit system will reveal the degree to which congestion delays actually are a serious matter to different commercial users of the waterways.

Facilitating Transactions and Disseminating Information

In nascent markets, it may be years before an equilibrium price is established and becomes common knowledge. Public agencies may take on the role of administering an auction (as with SO_2 allowances), setting offer prices (as with California Water Bank and the Edwards Aquifer land-fallowing program), or collecting and disseminating transaction data (as with the U.S. Environmental Protection Agency's Internet database on SO_2 allowance trades). Eventually, private brokers enter to provide services; such as with the SO_2 allowance market and in a few active western U.S. water markets. Confidentiality regarding prices and other transaction details sometimes is required by would-be traders as a condition for market participation. This is the case for the U.S. tradable permit program in ozone depleting chemicals (Tietenberg, 2000) and for water trading among irrigators in California's Central Valley who do not want price levels disclosed.

Lack of information on the value of the tradable right is a hurdle that can be overcome by public reporting of transactions. Public agency leadership in establishing real-time information channels for navigation users to trade permits would be essential to getting a trading program started. In addition, trading rules and reporting requirements should limit transaction costs to those necessary to account for trade-related information costs and externalities (if any). "Only emissions trading programs with low transaction costs have succeeded. "(United Nations, 1998, p. 24). A low-cost trading mechanism is particularly important in the context of trading lockage permits, as the gains from trade are likely to be modest, thus dampening trading by any notable costs and inconveniences.

Assuring Adequate Monitoring/Enforcement

A trading program quickly becomes meaningless if users can exceed use limits with impunity and so have little incentive to trade. Although the SO_2 allowance program involves continuous monitoring of emissions, stiff fines for noncompliance, and public access to data on trades, prices, and emissions, loopholes in groundwater pumping enforcement allow water users to expand their use beyond permitted levels, undercutting market development (Anderson and Snyder, 1997). Enforcement concerns have arisen in Alaska halibut and sablefish fisheries as the budget for monitoring and enforcement has been less than one-fourth of that requested and key staff positions have gone unfilled (NRC, 1999).

A meaningful lockage permit trading program will require monitoring as barges enter the locks to assure that they are indeed next in the queue, given assigned permits and any trades that have occurred. Fines or other penalties for violating established trading rules and rules for use of locks (such as allowable time in the lock) and moving through queues need to be established at levels that are high enough to discourage violations.

Documenting Welfare Gains

Documenting of the welfare gains from permit trading programs requires a well-defined baseline of comparison. In principle, one seeks to compare net benefits from resource use with and without the trading policy. In practice, this requires construction of a counterfactual — what would have occurred without the program? One could hypothesize a baseline in which command-and-control policies were used to achieve the same level of resource use and environmental quality that a tradable permit program achieves. Stavins (2000) uses this type of baseline to estimate abatement cost savings for the SO_2 allowance program.

Politically, adoption of a tradable lockage permit program requires an expectation of net benefits for relevant constituencies. In the case of navigation on the Upper Mississippi, benefits to navigation interests primarily would be in the form of reduced congestion delays. However, perceptions regarding the benefits hinge upon what stakeholders believe would occur in the absence of a permit program. If such a program is viewed as a means to live with the current lock infrastructure, given that no lock extensions are forthcoming, a permit program may be welcomed. If the program is viewed as delaying, or weakening the case for, lock extensions that otherwise would be forthcoming, it will be bitterly opposed. During the era when federal largesse was still available to develop additional water supplies, tradable water permits were an anathema to water users. Once that era had clearly ended, trading was an acceptable means of managing a finite resource.

Implications from Other Tradable Permit Programs

The advent of tradable permits signals a paradigm shift to a new era of declining subsidies, environmental constraints on customary use patterns, and resource transfers—changes often bitterly resisted by resource users and those businesses and communities linked to them. Some attempts to develop tradable permits may cycle through many rounds of public debate over rules for trading and who should get permits. Once trading begins, new litigation and ambiguities in the trading rules take time to be resolved.

A mature market with routine transactions and least-cost attainment of policy goals is the ideal outcome of a tradable permit program, an elusive outcome if the economic and political impetus for trading has not ripened sufficiently (cf. Colby, 2000). Consequently, those advocating a tradable permit approach do well to foster realistic expectations about how long it may take to establish a working program.

In some programs (fisheries ITQs, SO^2 allowances), early trades occurred primarily within a firm or parent corporation, with interfirm trading taking much longer to occur because of greater uncertainty about the reliability of the trading partner and higher information and negotiating costs. This phenomenon may be observed with tradable lockage permits as well and could limit efficiency gains from trading.

One of the advantages of tradable permits is the potential for environmental advocates to "put their money where their mouth is" by acquiring and permanently retiring permits, thus decreasing overall use of the resource. Such transactions are common with SO^2 allowances and water permits and might provide a vehicle for environmental advocates to directly influence navigation use of specific reaches of the UMR–IWW system. However, environmental acquisitions would be pointless if navigation interests could acquire additional lockage permits upon request and at zero cost.

In the case of inland waterway navigation (as with fisheries and water use), there are noncommercial users of the resource who must be considered when tradable permits are developed. Should recreational boaters be required to have a lockage permit, or can they better be accommodated by implementing standardized rules for queuing recreational boats at the locks? Noncommercial constituencies, when not adequately considered, create political and legal obstacles to a successful tradable permit program. Recreational fishing interests in Alaska still are effective gadflies in the fisheries ITQ programs. Noncommercial boaters are actively litigating river-running permit systems in national parks that do not address their interests. When noncommercial interests are neglected, there often are repercussions for the commercial permit-trading program.

Tradable permit programs are directly linked to the politics of public subsidies to resource users. Naturally, resource users strongly prefer new infrastructure development, paid for by tax dollars, to a tradable permit program. They will vigorously resist tradable permits and other nonstructural measures if federally subsidized improvements are viewed as still possible.

Trading rules that avoid imposing unnecessary costs help jumpstart the market, as does low-cost access to information on trading activity and prices during the interim period before the market becomes active enough to attract private brokerage and information services.

Finally, a powerful incentive is needed to achieve any notable policy change. A strong external mandate—a court ruling, an act of Congress, an international treaty—makes use limits and tradable rights more palatable. State and local officials can then blame the higher authority for the disruption of customary resource use and lobby heavily for compensation, creating new inefficiencies if they are successful.

REFERENCES

Anderson, T. L., and P. Snyder. 1997. Water Markets: Priming the Invisible Pump. Washington, D.C.: Cato Institute.

Colby, B. G. 2000. Cap-and-trade policy challenges: A tale of three markets. Land Economics 72: 638-658.

National Research Council. 1999. Sharing the Fish: Toward a National Policy for Individual Fishing Quotas. Washington D.C.: National Academy Press.

Stavins, R. N. 2000. Market based environmental policies, in Public Policies for Environmental Protection. P. R. Portnoy and R. N. Stavins, eds. Washington D.C.: Resources for the Future.

Tietenberg, T. H. 2000. Tradable permit approaches to pollution control: Faustian bargain or paradise regained? In Property Rights, Economics and the Environment. M.D. Kaplowitz, ed. Stanford, Connecticut: JAI Press.

Appendix D

Biographical Sketches
Committee to Review the Upper Mississippi River–Illinois Waterway Navigation System Feasibility Study

Lester Lave, ***Chair,*** (IOM) Harry B. and James H. Higgins Professor of Economics and University Professor, Carnegie Mellon University. Research interests include applied economics and public policy, safety goals for dams and other structures, quantitative risk assessment. Member, Science Advisory Board, U.S. EPA. Former president, Society for Risk Analysis. Ph.D., Harvard; B.A., Reed College.

Phillip Baumel, Professor of Economics and Charles F. Curtiss Distinguished Professor in agriculture, Iowa State University. Research interests are transportation and marketing of agricultural outputs and inputs and rural transportation problems. Ph.D., Iowa State University; M.S., and B.S., Ohio State University.

Kenneth D. Boyer, Professor of Economics at Michigan State University. Author of *Principles of Transportation Economics*. Former Chair, American Economic Association's Transportation/Public Utilities Group; member, Midwestern Economic Association, Secretary/Treasurer of the Industrial Organization Society. Former member of TRB Committee on Public Policy for Surface Freight Transportation. Current member, TRB Committee on Truck Size and Weights. M.A., Ph.D, University of Michigan; B.A., Amherst College.

Michael S. Bronzini, Sidney O. Dewberry Chair in Civil, Environmental and Infrastructure Engineering, George Mason University. Previously Director, Center for Transportation Analysis, Oak Ridge National Laboratories; Professor and Head of the Department of Civil Engineering, Pennsylvania State University; faculty member, University of Tennessee, Georgia Tech. Registered Professional Engineer. Former Chair, TRB Committee on Landside Access to Ports; former member, TRB Committee on Policy Options for Intermodal Freight Transportation. M.S., Ph.D., Pennsylvania State University; B.S., Stanford University.

Kenneth L. Casavant, Professor of Agricultural Economics, Washington State University. Research interests include transportation policy, agricultural marketing. Member, Board of Directors, Washington State Transportation-Policy Plan; American Agricultural Economics

Association; Regional Science Association. Former member, Washington State Governor's Joint Natural Resources Cabinet, Pacific Northwest Power Planning Council. Ph.D., Washington State University; B.S., M.S., North Dakota State University.

Bonnie G. Colby, Professor, Department of Agricultural and Resource Economics, University of Arizona. Research interests include natural resource economics, public policy, and water allocation. Member, American Agricultural Economics Association, American Water Resources Association, Association of Environmental and Resource Economists. Former member, President's Council on Sustainable Development Western Regional Team, NRC Committee on Western Water Management. M.S., Ph.D., University of Wisconsin; B.S., University of California, Davis.

Jonathan P. Deason, Lead Professor, Environmental and Energy Management Program, Department of Engineering Management and System Engineering, George Washington University. Previously Vice President, American Road and Transportation Builders Association (1994–1996); U.S. Department of the Interior (1986–1994, 1978–1984); Office of the Assistant Secretary of the Army-Civil Works (1984–1986); U.S. Army Corps of Engineers (1970–1978). Fellow, American Society of Civil Engineers; member, American Water Resources Association. Former member, Executive Committee, Renewable Natural Resources Foundation. Registered Professional Engineer. Ph.D., University of Virginia; M.S., Johns Hopkins University; M.B.A., Golden Gate University; B.S., U.S. Military Academy.

José A. Gómez-Ibáñez, Derek C. Bok Professor of Public Policy and Urban Planning, John F. Kennedy School of Government, Harvard University. Research interests are transportation policy, urban development, privatization, and regulation of infrastructure. Member, American Economic Association. Former Chair, TRB Committee on Public Policy for Surface Freight Transportation. A.B., M.P.P., Ph.D., Harvard University.

Delon Hampton (NAE), President and Chief Executive Officer, Delon Hampton & Associates, Chartered. Formerly Professor of Civil Engineering, Howard University; Research Engineer, Illinois Institute of Technology; Assistant Professor, Kansas State University. Professional practice is in engineering design and construction management. President, American Society of Civil Engineers. Associate Member, Board of Governors, American Public Transit Association; TRB Executive Committee. M.S.C.E., Ph.D., Purdue University; B.S.C.E., University of Illinois.

Edwin E. Herricks, Professor of Environmental Biology, Department of Civil and Environmental Engineering, and Departmental Affiliate, Department of Natural Resources and Environmental Sciences, University of Illinois. Research interests include methods of improving environmental decision-making, ecologically relevant engineering design. Member, American Water Research Association, American Society of Civil Engineers. Ph.D. (biology), Virginia Polytechnic Institute and State University; M.S., Johns Hopkins University; B.A., University of Kansas.

David H. Moreau, Professor and Chair, Department of City and Regional Planning, University of North Carolina. Research interests include planning, finance, and evaluation of water resources programs. Chairman, North Carolina Environmental Management Commission. Ph.D., Harvard University, M.S.C.E., North Carolina State University, B.S.C.E., Mississippi State University.